# 1,3-Bisphenylguanidine Based Monomeric Iridescents for Aberrant Illuminal Instantiations

Julian Princeps

1,3-Bisphenylguanidine Based Monomeric Iridescents for Aberrant Illuminal Instantiations

First Edition October 2025



Written by Julian Princeps

## TABLE OF CONTENTS

## LIST OF TABLES

## LIST OF FIGURES

## LIST OF SYMBOLS, ABBREVIATIONS AND NOMENCLATURE

| | |
|---|---|
| $\alpha$ | Absorption coefficient (In linear optics) |
| $\alpha$ | Polarizability (In nonlinear optics) |
| C | Actual concentration |
| a.u. | Atomic units |
| $E_g$ | Band gap energy |
| BBO | Beta barium borate |
| B3LYP | Becke's three parameters |
| $\eta$ | Conversion efficiency |
| CC | Coupled cluster |
| DFT | Density functional theory |
| DTA | Differential thermal analysis |
| $\mu$ | Dipole moment |
| $d_{eff}$ | Effective nonlinear coefficient |
| **E** | Electric field |
| EA | Electron affinity |
| e.s.u. | Electrostatic units |
| $C^*$ | Equilibrium concentration |
| $\beta$ | First order hyperpolarizability |
| FTIR | Fourier transform infrared |
| FT-Raman | Fourier transform Raman |
| $\Delta G_V$ | Free energy change per unit volume |
| GdCOB | Gadolinium calcium oxoborate |
| $\Delta G$ | Gibbs free energy change |
| HF | Hartree-Fock theory |
| ISE | Indentation size effect |
| IP | Ionization potential |
| IR | Infrared |

| | |
|---|---|
| LAP | L-arginine phosphate |
| LDT | Laser damage threshold |
| $LiNbO_3$ | Lithium niobate |
| $\chi$, $\chi^{(1)}$ | Linear susceptibility |
| LDA | Local density approximation |
| LUMO | Lower unoccupied molecular orbit |
| V | Molar volume of the crystal |
| MP | Moller-Plesset |
| DPGN | N,N'-Diphenylguanidinium nitrate |
| DPGTM | N,N'-Diphenylguanidinium hydrogen tartrate monohydrate |
| DPGP | N,N'-Diphenylguanidinium dihydrogen phosphite |
| DAST | 4-N,N-dimethylamino-4'-N'-methyl-stilbazolium tosylate |
| NLO | Nonlinear optics |
| $\chi^{(2)}$, $\chi^{(3)}$ | Nonlinear susceptibility |
| $\varepsilon_0$ | Permittivity of free space |
| $\Delta K$ | Phase–mismatching |
| h | Planck's constant |
| **P** | Polarization |
| KBr | Potassium bromide |
| $KB_5$ | Potassium pentaborate |
| KDP | Potassium dihydrogen ortho-phosphate |
| $KHo(WO_4)_2$ | Potassium holomium double tungstate |
| KTP | Potassium titanyl phosphate |
| PXRD | Powder X – ray diffraction |
| QCISD | Quadratic configuration interaction |
| SHG | Second harmonic generation |
| SCF | Self-consistent field |
| SXRD | Single Crystal X – ray diffraction |
| S | Supersaturation |
| TG | Thermogravimetric analysis |

| | |
|---|---|
| T | Transmittance |
| TS | Transition state |
| UV | Ultraviolet |
| c | Velocity of light |
| $\lambda$ | Wavelength |
| XRD | X – ray diffraction |
| YCOB | Yttrium calcium oxyborate |

# CHAPTER – I

# INTRODUCTION TO CRYSTAL GROWTH TECHNIQUES, NONLINEAR OPTICS AND DENSITY FUNCTIONAL THEORY

## 1.1 INTRODUCTION

Crystals form the backbone of modern technological advancement, playing a vital role in science, engineering, and technology. The study and growth of crystals have evolved into a precise scientific discipline, driven by the wide range of applications of single crystals. These materials serve as fundamental building blocks for innovations in fields such as medicine, electronics, defense, space science, and communications.

Single crystals are characterized by a periodic atomic arrangement in three dimensions, where the environment of each atom is uniform relative to similar atoms. Ideal crystals exhibit perfect uniformity, but real crystals are finite and often contain defects. Nevertheless, the regularity of single crystals enables the transmission of electromagnetic waves with minimal scattering, which is critical in many technological applications. Over the past several decades, the ability to grow tailor-made crystals has propelled significant advancements in electronics, fiber optics, and laser technologies. The precise control over crystal structure and quality has made it possible to develop highly specialized materials that meet the demands of modern scientific and industrial applications.

It is clearly more difficult to develop single crystal than poly-crystalline material and extra effort is justified because of the outstanding advantages of single crystals (Laudise, 1970). The chief advantages of single crystals are the anisotropy, uniformity of composition and the absence of boundaries between individual grains. The study of anisotropy of the properties of crystals requires specimens with different orientations of the same single crystal. This can be easily made from large size crystal. The strong influence of single crystals in the present day technology is evident from the recent advancements in the above mentioned fields. Hence, in order to achieve high performance from the device, good quality single crystals are needed.

With the invention of nonlinear optical (NLO) properties in single crystals, tunable lasers have been realized. As there is a wide market for solid state devices in the fields of computers, telecommunication etc., efforts have been made in recent years on producing larger single crystals (Brice, 1986). Growing crystals of desired size and shape is still considered as an art. Everyday new materials are being grown and new devices are coming out from these crystals. In recent days, crystals are produced artificially to satisfy the needs of science, technology and ornamental purposes. The ability to grow high quality crystals has become an essential criterion for the competitiveness of world. Crystal growth specialists have been moved from the periphery to the center of the materials based technology.

A variety of crystals are needed to meet some very important gaps in conventional production engineering. Many types of crystals find applications in lasers, nonlinear optics (NLO), light emitting diodes (LED), pyroelectric detectors and optical components for communication. Semiconductor lasers, piezoelectric, pyroelectric, ferroelectric, scintillating and infrared sensitive crystals are a part of several solid state devices (Bailey et. al., 1991; Nalwa and Miyata, 1997a).

New materials are the basis for solid state research and device technology. On contrary to what many believe, new materials are not usually discovered by device engineers and solid state theoreticians; they are mostly discovered by crystal chemists who are crystal growers. Some physical phenomena are only exhibited in single crystals and can only be studied and understood in single crystals. Thus the crystal grower develops a proficiency in relating structure, bonding and other physiochemical considerations to properties of interest that is in a key position in determining the direction and success of solid state research and ultimately technology. Growth of single crystals and their characterization towards device fabrication have assumed great impetus due to their importance for both academic as well as applied research.

## 1.2 NUCLEATION

Nucleation is an important phenomenon in crystal growth and is the precursor of the overall crystallization process. Nucleation is the process of generating within a metastable mother phase, the initial fragments of a new and more stable phase capable of developing spontaneously into gross fragments of the stable phase. Nucleation is consequently a study of the initial stages of the kinetics of such transformations (Santhanaraghavan and Ramasamy, 2000).

Nucleation may occur spontaneously or it may be induced artificially. These two cases are referred to as homogeneous and heterogeneous nucleations respectively. Both these nucleations are called primary nucleation and occur in systems that do not contain crystalline matter. On the other hand, nuclei are often generated in the vicinity of crystals present in the supersaturated system. This phenomenon is referred to as secondary nucleation (Mullin, 1972).

The growth of crystals from solutions can occur if some degree of supersaturation or supercooling has first been achieved in the system. There are three steps involved in the crystallization, which are

(i). achievement of supersaturation or super cooling,

(ii). formation of crystal nuclei and

(iii). successive growth of crystals to get distinct faces.

### 1.2.1 Ostwald's Contributions

While the concept of a definite supersolubility is contained in the earliest writings on the crystal growth subject, Ostwald appears to be the first to explain the relationship between supersaturation and spontaneous crystallization. The relationship between the concentration and temperature is schematically shown in figure1.1 (Meirs and Issac, 1987).

Meir's solubility diagram consists of three zones, which are termed as region I, II and III respectively. Region I corresponds to the undersaturated zone, where crystallization is not possible. This region is thermodynamically stable. The region II between the supersolubility curve and solubility curve is called as metastable zone, where spontaneous crystallization is not possible but seeded growth can be initiated in this region. In the region III, spontaneous nucleation is more probable, called as unstable or labile zone.

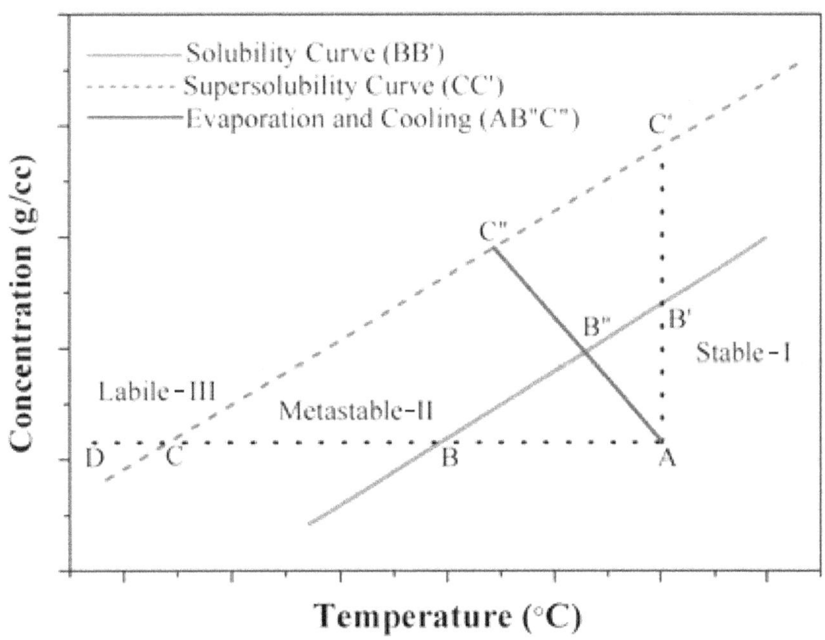


AB'C' – Evaporation; ABC – Cooling.

**Figure 1.1 Meir's solubility diagram showing different levels of saturation**

### 1.2.2 Supersaturation

A solution in which the concentration of the solute exceeds that of the equilibrium condition at a given temperature is known as supersaturated solution. A supersaturated solution is thermodynamically unstable. Supersaturation required for crystallization process can be achieved by (i). slow evaporation, (ii). slow cooling and (iii). by adding any external impurity. Supersaturation achieved by slow cooling is the best method to grow bulk size crystals by solution technique. In this case, crystallization takes place by lowering the temperature of the solution under a controlled cooling rate. Supersaturation can be achieved by slow evaporation of the solvent for materials, which are having very small temperature coefficient of solubility. Here crystallization can take place at a constant temperature. Supersaturation can also be achieved by the addition of some impurities. The growth proceeds by the reduction of solubility of the solute due to the presence of impurity. The selection of suitable growth methods depends mainly on the shape of the solubility curve.

The degree of supersaturation of a solution can be expressed by the ratio :

$$S = \frac{C}{C^*} \qquad (1.1)$$

where, $C$ the actual concentration of the solution and $C^*$ the equilibrium saturation concentration of the solution at a given temperature. Thus, for a saturated solution, $S = 1$, $S < 1$ denotes under saturation and $S > 1$ indicates supersaturation.

The concentration driving force ΔC is given by

$$\Delta C = C - C^* \qquad (1.2)$$

The relative supersaturation is defined as,

$$\delta = \frac{C - C^*}{C^*} \qquad (1.3)$$

$$\delta = \frac{C}{C^*} - 1 \qquad (1.4)$$

$$\delta = S - 1 \qquad (1.5)$$

Crystallization process is initiated by the formation of embryos or nuclei with number of micro size solid particles present in the solution, termed as centers of crystallization. The mechanism of formation of such embryos is the simple collision process of a single molecule $A_1$ with a cluster $A_{i-1}$ consisting of (i -1) molecules and thus giving rise to a cluster $A_i$; i.e.,

$$A_{i-1} + A_1 \rightarrow A_i \qquad (1.6)$$

The total Gibb's free energy change, ΔG of the embryo between the two phases associated with this process is given as

$$\Delta G = \Delta G_S + \Delta G_V \qquad (1.7)$$

where $\Delta G_S$ is the surface free energy change and $\Delta G_V$ is the volume free energy change

For the spherical nucleus of radius, *r*

$$\Delta G = 4\pi r^2 \gamma + \frac{4}{3}\pi r^3 \Delta G_V \qquad (1.8)$$

where γ the interfacial tension and $\Delta G_V$ the free energy change per unit volume and is a negative quantity. The quantities ΔG, $\Delta G_S$ and $\Delta G_V$ are represented in figure 1.2. Since the surface free energy increases with $r^2$ and volume free energy decreases with $r^3$, the total net free energy change increases with increase in size and attains a critical size after which it decreases.

At critical condition, the free energy of formation obeys the condition,

$$\frac{d\Delta G}{dr} = 0 \qquad (1.9)$$

Hence the radius of the critical nucleus is expressed as

$$r^* = \frac{-2\gamma}{\Delta G_v} \qquad (1.10)$$

$$\Delta G_v = \frac{-kT\ 1nS}{V} \qquad (1.11)$$

V = molar volume of the crystal

k = Boltzmann constant

T = Temperature of the solution in K

Hence, the critical free energy barrier

$$\Delta G = \frac{16\pi\gamma^3\ V^3}{3\mu^2} \qquad (1.12)$$

The number of molecules in the critical nucleus is expressed as

$$i^* = \frac{4\pi\ (\gamma^*)^3}{3V} \qquad (1.13)$$

The crucial parameter between a growing crystal and the surrounding mother solution is the interfacial tension ($\gamma$). This complex parameter can be determined by conducting the nucleation experiments.

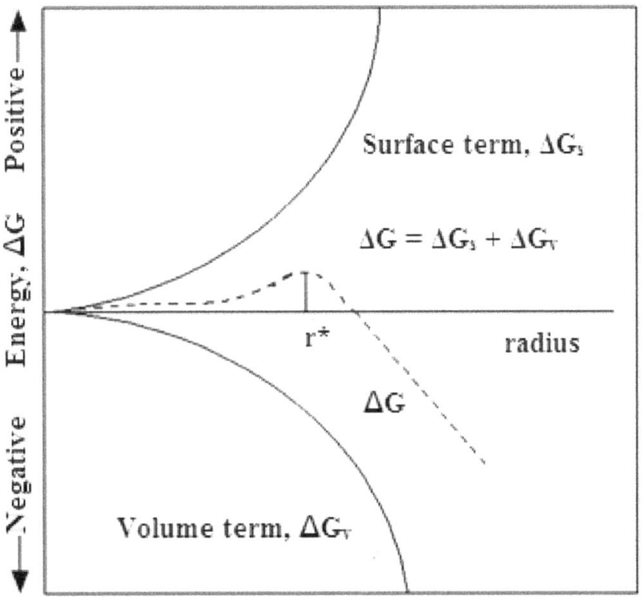


**Figure 1.2 Change in free energy due to the formation of nucleus**

## 1.3 METHODS OF CRYSTAL GROWTH

Crystal growth is a controlled phase transformation to a solid phase, either from solid or liquid or gaseous phases. The process of producing single crystals, from homogeneous media at the atomic level with directional properties, attracts more attention than any other processes. The methods of growing crystals are very wide and are mainly dictated by the characteristics of the material and its size (Buckley, 1951; Mullin, 1976). The phase transformations involved in crystal growth methods are given below,

Growth from solid ⟶ Solid–solid phase transformation

Growth from liquid ⟶ Liquid–solid phase transformation

Growth from vapour ⟶ Vapour–solid phase transformation

The conversion of a polycrystalline material into a single crystal by causing the grain boundaries to be swept through and pushed out of the crystal in the solid-growth of crystals (Mullin, 1972). An efficient process is the one which produces large size single crystals for applications at minimum cost. The selection of growth method is vital because it suggests the possible impurity and other defect concentrations. By choosing the best method to grow single crystals, we can able to grow bulk crystals but however it depends on the characteristics of the material and requirements.

## 1.4 GROWTH FROM SOLUTION

Among the various methods of growing single crystals, solution growth occupies a prominent role owing to its versatility and simplicity. The growth of crystals from solution occurs close to equilibrium condition and hence bulk single crystals with clear perfection can be grown easily.

Growth of crystals from aqueous solution is a classical method. In this method, the crystals are grown at a temperature well below the melting point. Materials, which have high solubility and have variation in solubility with temperature can be grown easily by solution method. The slow growth rate and selection of suitable solvent are the disadvantages of this method. In this method, solvent must have high solubility, good solubility gradient, low volatility, density less than that of a bulk solute and low viscosity.

The various solution growth technique are,

(i). hydrothermal growth,

(ii). gel growth,

(iii). high temperature solution growth and

(iv). low temperature solution growth.

Out of the various types of solution growth, the low temperature solution growth is majorly focused, since growth is carried out near room temperature, the structural imperfections in solution grown crystals are relatively low (Brice,1972).

The low temperature solution growth method can be subdivided into the following types :

(i). slow cooling technique,

(ii). slow evaporation technique and

(iii). temperature gradient technique.

### 1.4.1 Slow Cooling Technique

Among the other solution growth techniques, slow cooling is the best technique to grow bulk single crystals from solution. A solution that is in equilibrium with the solid phase is said to be saturated with respect to that solid. The supersaturation of the solution achieved by systematic cooling requires the volume of the crystallizer and the amount of substance in it. Hence, volume of the crystallizer is selected based upon the desired size of the crystal and solubility of the substance for a particular temperature. The temperature at which such crystallization can begin is usually in the range 45 – 70°C and lower limit of cooling is the room temperature. Recently, single crystal of nonlinear optical tetra L-lysine alanine mono hydrochloride dihydrate was grown by slow cooling technique (Siva Shankar et. al., 2008).

### 1.4.2 Slow Evaporation Technique

In this process, the temperature of the solution is not changed, but the solvent in the solution is allowed to evaporate slowly. Since the solvent evaporates, concentration of solute in the solution increases, supersaturation is achieved. In contrast to the cooling method, in which the total mass of the system remains constant, the solution loses particles, which are weakly bound to other components in the solvent evaporation method and therefore, the volume of the solution decreases. Since the vapor pressure of the solvent is higher than the vapor pressure of the solute, the solvent evaporates more rapidly and the solution becomes supersaturated. Typical growth conditions involve temperature stabilization to about very few °C and the rate of evaporation of a few $mm^3/hr$. This method can effectively be used for materials having very low temperature coefficient of solubility. Large size single crystals of N,N'- diphenylguanidinium nitrate and glycine sodium nitrate were grown under slow evaporation technique (Saravana Kumar and Murugakoothan, 2014; Narayan Bhat and Dharmaprakash, 2002).

### 1.4.3 Temperature Gradient Technique

Temperature gradient technique involves the transport of materials from a hot region containing the source material to be grown to a cooler region, where the solution is supersaturated and the crystal grows. The main advantages of this technique are,

(i). crystal grows at a fixed temperature,

(ii). insensitive to changes in temperature provided both the source and the growing crystal undergo the same change and

(iii). economy of solvent and solute.

On the other hand, changes in the small temperature difference between the source and the growing crystal have a large effect on the growth rate. In this method, the possibility of growing bulk crystals is admissible. The temperature gradient technique is very effectively employed in the high temperature solution growth (Flux Growth). Large crystal of Beta barium borate (BBO) was grown by Eugene et. al., (2005). Majchrowski et. al., (2004) reported the growth of Potassium holomium double tungstate ($KHo(WO_4)_2$) single crystal by temperature gradient top seeded solution growth technique.

## 1.5 LOW TEMPERATURE SOLUTION GROWTH METHODOLOGY

The method of crystal growth from low temperature aqueous solution is extremely popular in the production of many technologically important crystals. The various parameters with regard to low temperature solution growth methodology are discussed below.

### 1.5.1 Solvent Selection

A given solute dissolves in different solvents with different solubility. In addition to the high purity starting materials, solution growth requires a good solvent. The best solvent should satisfy the following conditions,

(i). high solubility for the given solute,
(ii). high positive temperature coefficient of solubility,
(iii). low viscosity,
(iv). low volatility,
(v). cost effective,
(vi). easy availability,
(vii). low toxicity and
(viii). density less than that of the bulk solute.

Apart from all the above characteristics an ideal solvent should yield a prismatic habit in the crystals (Pamplin, 1979).

A mixture of two or more solvents will occasionally be found to possess the best properties for a particular crystallization purpose. A second liquid is sometimes added to a solution to reduce the solubility of the solute, cause its precipitation/crystallization and maximize the yield of product. Some of the main points that should be taken into account when choosing a solvent for a crystallization process include the following factors mentioned above. Also, the solute to be crystallized should be readily soluble in the solvent. It should also be easily deposited from the solution in the desired crystalline form after cooling, evaporation etc.

Solvents may be classified into three categories viz. polar, dipolar and non-polar. Usually, a polar solute is more soluble in polar solvent than in non-polar solvent. Similarly, a non-polar solute is more soluble in non-polar solvent than in polar solvent. However, a close chemical similarity between the solute and solvent should be avoided, because their mutual solubility is high in all the ways which leads to difficult or uneconomical crystallization process. It is very important to note that the crystal habit can often be changed by the solvent.

### 1.5.2 Determination of Solubility

The solubility data at various temperatures are essential to determine the level of supersaturation. Hence, the solubility of the solute in the chosen solvent must be determined before starting the growth process (Christian, 1990). If the solubility is too high, it is difficult to grow bulk single crystals and too low solubility restricts the size and growth rate of the crystals. Neither a flat nor a steep solubility curve will enable the growth of bulk crystals from solution. If the solubility gradient is very small, slow evaporation of the solvent is the other option for crystal growth to maintain supersaturation in the solution.

The solute is dissolved in the solvent maintained at a constant temperature with continuous stirring to determine the solubility of the solute. On reaching saturation, the equilibrium concentration of the solute can be determined gravimetrically. The solubility curve can then be plotted between the amount of solute dissolved and temperature by repeating the measurements for different temperatures.

### 1.5.3 Crystal Habit

Mostly, all crystals grow approximately at equivalent rates in all directions. The imperfections become isolated into defective regions surrounded by large volumes of high perfection, when the crystal grows with a bulk habit. Furthermore, during the growth where dislocations and other defects propagate, they do so from the nucleus or seed along specific directions into the bulk of the crystal. In the crystals, which grow as needles or plates, the growth dislocations propagate along the principal growth directions and the crystals remain imperfect (Pamplin, 1979). Needle like crystals have very limited applications and plate like crystals need to be favorably oriented.

Changes of habit in such crystals, which naturally grow as needles or plates can be achieved by any one of the following ways :

(i). changing the temperature of growth,

(ii). changing the pH of the solution,

(iii). adding a habit modifying agent and

(iv). changing the solvent.

Achievement of bulk crystal growth is of great industrial importance where such morphological changes are induced during crystallization to yield crystal with better perfection and packing characteristics.

### 1.5.4 Solution Preparation

For solution preparation, it is essential to have the solubility data of the material at different temperatures. Sintered glass filters and papers with different pore sizes are used for solution filtration. The clear solution, saturated at the desired temperature is taken in a growth vessel. For crystal growth by slow cooling, the vessel is sealed to prevent the solvent evaporation. Solvent evaporation at constant temperature can be achieved by providing a controlled vapour leak. A small crystal suspended in the solution is used to test the saturation. By varying the temperature, a situation is established where neither the occurrence of growth nor dissolution. The test seed is replaced with a good quality seed. All unwanted nuclei and the surface damage on the seed are removed by dissolving at a temperature above the saturation point. Growth is initiated after lowering the temperature to the equilibrium saturation. Solvent evaporation can also be helpful in initiating the crystal growth.

### 1.5.5 Seed Preparation

The quality of the grown crystal depends on the quality of the seed crystal used. Seed crystals can be obtained by spontaneous nucleation in the labile region from the supersaturated solution. The seed used to grow a large uniform crystal must be a single crystal free of inclusions, cracks, block boundaries, sharp cleaved edges, twinning and any other obvious defects. It should be of minimum size with at least little hardness, compatible with other requirements. When larger crystals of the same material are already available, they can be cut in the required orientation to fabricate the seed crystal. Since the growth rate of the crystal depends on the crystallographic orientation, the seed crystal must be cut in such a way that it has larger cross-section in the fast growing direction.

## 1.6 INTRODUCTION TO NONLINEAR OPTICS

Nonlinear optics (NLO) is the study of the interaction of intense electromagnetic field with materials to produce modified field that is different from the input field in phase, frequency, amplitude and polarization. Nonlinear optics holds the key for the future technology of photonics which would utilize optical frequency conversion, optical signal processing and image processing using ultra-fast laser pulses. The advantages of photonics as compared to electronics are the gain in speed, band width and interconnectivity. The most commonly used second order nonlinear optical effects are frequency doubling and electro-optic modulation. For second order nonlinear optical effects, which also include sum and difference frequency generations, optical parametric amplification and in most cases also photorefractive effects, noncentrosymmetry is required.

In the last few years, there have been intensive research efforts in exploring and developing efficient second order nonlinear optical materials for various applications, such as electro-optic switching or modulation for telecommunication and frequency conversion for laser sources at newer wavelengths. Such materials can be combined with inexpensive diode lasers for optical parameter generation of a broad range of optical frequencies. In addition, acentric crystals, optimized for optical frequency conversion, have a large potential for realization of low-power optical switching due to cascading of second order nonlinearities. Crystalline solids are often regarded as the best choice of material form for practical use as frequency converters because of the excellent long-term orientational stability and extremely high density of active components. In order to become a potential candidate for practical applications of this kind, the materials need to exhibit large second order optical nonlinearities.

### 1.6.1 Nonlinear Optical Effect

The phenomenon of nonlinear optical effect is understood by considering the process of the dielectric material at the atomic level to the electric fields of an intense light beam. In the case of linear optics, the induced polarization (**P**) is directly proportional to the applied electric field (**E**),

$$\mathbf{P} = \varepsilon_0 \, \chi^{(1)} \, \mathbf{E} \tag{1.14}$$

where the constant of proportionality, $\chi^{(1)}$ the linear susceptibility and $\varepsilon_0$ the permittivity of free space.

At high fields, polarization becomes independent of the field and the susceptibility becomes field dependent. Therefore, the nonlinear response can often be described by generalizing Eq. (1.14) by expressing the polarization **P**(t) as a power series in the field strength **E**(t) (Boyd, 2007).

$$\mathbf{P}(t) = \varepsilon_0 \left[\chi^{(1)}\, \mathbf{E}(t) + \chi^{(2)}\, \mathbf{E.E}\,(t) + \chi^{(3)}\, \mathbf{E.E.E}\,(t) + \ldots\ldots\right] \tag{1.15}$$

The quantities $\chi^{(2)}$ and $\chi^{(3)}$ are known as the second and third-order nonlinear optical susceptibilities, respectively.

In centrosymmetry crystals, the terms of even powers of 'E' are absent. However, in noncentrosymmetric crystals, all the terms are present. By substituting the electric field, E (t) = $E_0$ cos ωt in Eq. (1.15) where ω is the angular frequency of the light (2πf), equation (1.15) then becomes,

$$\mathbf{P}(t) = \varepsilon_0 \left[\chi^{(1)}\, \mathbf{E}_0 \cos \omega t + \chi^{(2)}\, \mathbf{E}_0.\mathbf{E}_0 \cos^2 \omega t + \chi^{(3)}\, \mathbf{E}_0.\mathbf{E}_0.\mathbf{E}_0 \cos^3 \omega t + \ldots\ldots\right] \tag{1.16}$$

Trigonometric substitution can be made for the $\cos^2 \omega t$ and $\cos^3 \omega t$

$$\cos^2 \omega t = \frac{1}{2} + \frac{1}{2} \cos 2\omega t \tag{1.17}$$

$$\cos^3 \omega t = \frac{3}{4} \cos \omega t + \frac{1}{4} \cos 3\omega t \tag{1.18}$$

The equation (1.16) becomes

$$\mathbf{P}(t) = \varepsilon_0\, [\chi^{(1)}\mathbf{E}_0 \cos \omega t + \frac{1}{2}\, \chi^{(2)}\, \mathbf{E}_0.\mathbf{E}_0 + \frac{1}{2}\, \chi^{(2)}\, \mathbf{E}_0.\mathbf{E}_0 \cos 2\omega t + \frac{3}{4}\, \chi^{(3)}\, \mathbf{E}_0.\mathbf{E}_0.\mathbf{E}_0 \cos \omega t + \frac{1}{4}\chi^{(3)}\, \mathbf{E}_0.\mathbf{E}_0.\mathbf{E}_0 \cos 3\omega t + \ldots\ldots \tag{1.19}$$

The equation (1.19) contains a term with an order 2ω; one component of the charge polarization equation is at twice the fundamental frequency of the original laser beam. As the incident beam falls on the crystal, oscillating electric dipoles absorb light at a frequency (ω) and reradiate it at the same frequency (ω) and at twice the original frequency (2ω). The total energy of the beam is not altered much, but is now split between two components. The phenomenon of emission of radiation of twice the original frequency (2ω), the second harmonic of the original beam, is called second-harmonic generation (SHG) (Sauter et. al., 1996). Thus, the harmonic waves are generated from a NLO material when it is placed in an intense laser beam.

Now for centrosymmetric materials, the second harmonic term cannot occur but in case of noncentrosymmetric materials both the second and third harmonic terms are present. Second harmonic generation, therefore, cannot occur in an isotropic medium, such as liquid or gases. However the cubic term is substantially smaller than the second harmonic term and may be ignored.

Equation (1.19) reveals not two, but three output terms: in addition to the fundamental (ω) and second-harmonic (2ω) terms, a constant term $\frac{1}{2}\chi^{(2)}\,\mathbf{E_0}.\mathbf{E_0}$, which represents a DC offset or steady polarization. It also represents an optical rectification analogous to that of a diode rectifying an AC voltage. Indeed, this DC signal (although a minute quantity) can be detected across a nonlinear crystal by placing electrodes on either side of the crystal, and this signal is proportional to the intensity of the light beam passing through the crystal.

In most crystalline materials, the nonlinear polarizability medium depends on the direction of propagation, polarization of electric field and orientation of the optic axis of the crystal. Since the vectors **P** and **E** are not necessarily parallel in such crystalline materials, the coefficient $\chi$ must be treated as tensor. The relation for the second harmonic polarization as

$$\mathbf{P}_i^{(2)} = \varepsilon \sum_{jk} \chi_{ijk}^{(2)} \mathbf{E}_j.\mathbf{E}_k \tag{1.20}$$

where i, j, k represents the coordinates of x, y, z. Sometimes second order nonlinear optical coefficient ($d_{ijk}$) with dimension of inverse field is preferably used instead of $\chi$ because d is defined under crystallographic axes as opposed to $\chi$ which is defined under cartesian coordinates.

In second harmonic generation, the two input wavelengths are the same i.e., the second harmonic wave with frequency, $\omega_2 = 2\omega_1$ ($\omega_1$ - frequency of fundamental wave; $\omega_2$ - frequency of extraordinary wave) is produced. The refractive index of ordinary wave, ($n_1$) is defined by the phase velocity and wavelength of the medium. The energy of the polarized wave is transferred to the electromagnetic wave at frequency ($\omega_2$). The phase velocity and wavelength of this wave are determined by ($n_2$), the refractive index of the medium for double frequency.

Traditionally the usefulness of a nonlinear optical crystal has been evaluated in terms of materials' parameters that are directly related to optical frequency conversion, such as (i) conversion efficiency, (ii) optical damage threshold, (iii) birefringence, (iv) dispersion, (v) optical transparency range and (vi) optical homogeneity.

The vectors of input and generated beams are to be matched to obtain high conversion efficiency,

$$\Delta K = \frac{2\pi}{\lambda(n_1 - n_2)} \tag{1.21}$$

where, $\Delta K$ represents the phase mismatching. The phase mismatch can be obtained by angle tilting, temperature tilting or other methods. The usual technique for achieving phase matching in nonlinear crystals is birefringent phase matching, where one exploits birefringence to cancel the phase mismatch. This technique comes in many variations : *Type-I phase matching* means that, e.g., in sum frequency generation, the two fundamental beams have the same polarizations, perpendicular to that of the sum frequency wave. Conversely, in *type-II phase matching*, the two fundamental beams have different polarization directions; this can be appropriate when the birefringence is relatively strong (overcompensating the dispersion in a type-I scheme) and/or the phase velocity mismatch is small. The distinction between type-I and type-II similarly applies to frequency doubling, and to processes such as degenerate or nondegenerate parametric amplification. The different polarization arrangements can have various practical implications for the combination of several nonlinear conversion stages, or for intracavity frequency doubling.

Hence to choose a nonlinear optical crystal, for a frequency conversion process, the necessary criterion is to obtain high conversion efficiency. The conversion efficiency ($\eta$) is given by the following relation (Prasad and Williams, 1991),

$$\eta = PL^2 \left( \frac{d_{eff} \, \mathrm{Sin}\, \Delta K.L}{\Delta K.L} \right)^2 \qquad (1.22)$$

where $d_{eff}$ the effective nonlinear optical coefficient and P the input power density. For a given material and for a given frequency, the conversion efficiency is dependent on the square of the interaction length L, hence it is imperative that large crystals are needed. As a result of the interference from the input fundamental polarization wave and the driven harmonic polarization wave, the maximum useable length of the crystal is limited to a few microns. However, this problem could be circumvented by decreasing '$\Delta K$' to zero technically known as phase matching (Kar, 2012). In a chosen material and selected interaction geometry, conversion efficiency can be further enhanced by increasing the power density of the input beam. This can be physically achieved by focusing the beam onto the crystal, increasing the power densities to the tolerance limit, called the damage threshold.

## 1.7 CLASSIFICATION OF NONLINEAR OPTICAL CRYSTALS

Nonlinear optical materials will be the key elements for future photonic technologies based on the fact that photons are capable of processing information with the speed of light.

Studies on the materials for nonlinear optical applications can be divided into three areas,

(i). discovery of new nonlinear materials,

(ii). growth and perfection of promising NLO crystals of desired shape and size for device applications and

(iii). improving the property modification of NLO crystals.

For optical applications, a NLO material should have the following characteristics,

(i). wide transparency domain,

(ii). large nonlinear figure of merit for frequency conversion,

(ii). high laser damage threshold,

(iv). availability in large single crystal forms,

(v). wide phase matching angle,

(vi). ability to process into crystal and thin film form,

(vii). ease of fabrication,

(viii). nontoxicity and good environmental stability,

(ix). high mechanical strength and thermal stability and

(x). fast optical response time.

For exhibiting SHG in materials, there are two factors, which determine the existence or absence of efficient SHG. Firstly, and fundamentally, the material should crystallize in a non-centrosymmetric crystal structure. Secondly for maximum SHG efficiency, crystal should possess phase matching properties (i.e., the propagation speeds of the fundamental and harmonic waves should be identical in the crystal). A typical SHG active molecule must be non-centrosymmetric in nature. This symmetry requirement eliminates many materials from being SHG active and hence at the early stage of designing and synthesizing novel materials, one has to consider the ways of introducing non-centrosymmetry in the molecular structures.

Nonlinear optical materials are broadly classified into :

(i). Inorganic crystals,

(ii). Organic crystals and

(iii). Semi-organic crystals.

### 1.7.1 Inorganic NLO Crystals

Over the past few decades, the design and synthesis of noncentrosymmetric (NCS) and polar inorganic materials are of particular interest due to their potentially important functional properties, such as piezoelectricity, pyroelectricity, ferroelectricity and second-order nonlinear optical (NLO) activity. After the invention of lasers, research on nonlinear optical phenomena in crystalline materials was concentrated and found nonlinear effect in inorganic materials, such as quartz, potassium dihydrogen phosphate (KDP), lithium niobate

($LiNbO_3$), urea, potassium titanyl phosphate (KTP) etc., and semiconducting crystals. Most commercial materials are inorganic for nonlinear optical applications especially in high power lasers. Potassium dihydrogen phosphate is the most widely used crystals for second harmonic generation due to its wide transparency range, good birefringence and moderate optical damage density. It can no longer be used as it is little hygroscopic. After many attempts, good nonlinear borate crystals have been discovered with wide transparent range. The first nonlinear optical effect on borate series crystals was observed in potassium penta borate ($KB_5$) (Dewey et. al., 1975). As a consequence of this, many inorganic crystals have been grown and investigated. In recent years, newly developed nonlinear optical crystals, $GdCa_4O(BO_3)_3$ (GdCOB) and $YCa_4O(BO_3)_3$ (YCOB) have attracted much attention due to their promising optical properties (Chen et. al.,1989). The family of various borate crystals thus plays a very important role in the field of nonlinear optics. Halasyamani et.al., (1998) and his collaborators have reported a series of polar oxide inorganic materials and they synthesized a series of noncentrosymmetric and polar oxides with cations susceptible to second-order Jahn Teller (SOJT) distortions, specifically, octahedrally coordinated $d^0$ transition metals ($Ti^{4+}$, $V^{5+}$, $Mo^{6+}$, $W^{6+}$ etc.) and cations with stereoactive lone pairs ($I^{5+}$, $Te^{4+}$, $Se^{4+}$, $Sb^{3+}$ etc.), both of which are in asymmetric coordination environments. Although inorganic materials possess good mechanical and thermal stabilities, the optical nonlinearity is relatively lower than organic and semi-organic materials. The growth of inorganic crystals is usually more difficult and typically adopts high temperature growth technique.

### 1.7.2 Organic NLO Crystals

Organic materials are in increasing demand, as they are better candidates in optical modulation, second-harmonic generation (SHG), optical signal processing, optical switching, optical data storage devices etc. It is also used in phase conjugation and image reconstruction. Organic molecules containing conjugate systems have some advantages over inorganic materials because of the possibility of highly enhanced electronic nonlinear optical polarization responses. The basic structure of organic NLO materials is based on the $\pi$-bond system; due to the overlap of $\pi$-orbitals, delocalization of electronic charge distribution leads to a high mobility of the electron density. Functionalization of both ends of the $\pi$-bond system with appropriate electron donor and acceptor groups can enhance the asymmetric electronic distribution in both ground and excited states, leading to an increased optical nonlinearity (Zyss, 1994). Organic crystals show a complex range of phase behavior, photo and thermal stabilities, solubility and morphology. The rapid development of optical communication systems has led to a demand for nonlinear optical materials of high structural and optical quality. The most widely encountered crystals for this type of applications are Urea, 4′-(dimethylamino)-*N*-methyl-4-stilbazolium tosylate (DAST), meta-nitroaniline (m-NA), 3-methyl-4-nitropyridine-1-oxide (POM) etc., (Chemla and Zyss, 1987).

The advantages of organic nonlinear optical materials are :

(i). high second order nonlinear optical efficiency,

(ii). birefringence used for phase matching and

(iii). possibility to chemically engineer the molecular properties.

The superiority of organic NLO materials results from their versatility and possibility of tailoring them for a particular device application. Organic NLO materials have high nonlinear figure-of-merit for frequency conversion and fast optical response time as compared with inorganic NLO materials. The search for new materials is primarily focused on increasing the nonlinearity. With the progress in crystal growth technology, materials having attractive nonlinear properties are being discovered at rapid pace (Chemla and Zyss, 1987). However there are some drawbacks with organic NLO materials. A major obstacle to the development of many organic NLO materials and the exploitation of their full potential has considerable difficulty associated with the growth of large size high quality single crystals of these materials. As with many other organic solids, the intermolecular forces are comparatively weak, being predominantly van der Waals or permanent dipole-dipole interactions. This typically results in low melting point and relatively high vapour pressure. Mechanical properties are, in general, rather poor with most organic solids being relatively soft. This can have important consequences for the structural perfection of the crystals.

### 1.7.3 Semi-Organic NLO Crystals

The inherent limitations on the maximum attainable nonlinearity in inorganic materials and the moderate success in growing device grade organic single crystals have made scientists to adopt newer strategies. The obvious one is to develop hybrid inorganic-organic materials with little trade-off in their respective advantages. This new class of materials has come to be known as the semi-organics. The widest search for new compounds and crystals led to the development of many amino acid based semi organic single crystals. Semi organics are primarily classified into two categories, type-I and type-II. Type-I semi-organics are inorganic salts comprising of large conjugated organic molecules. The organic part in type-I are generally chiral compounds, like tartrates, oxalates and a host of amino acids. L-arginine phosphate monohydrate (LAP) is a typical example of this class. Type-II semi organics are coordination complexes of polarizable organics bonded to metal atoms. The metal atoms help in forming relatively strong coordinate bonds resulting in better thermal and mechanical stabilities. Zinc tris thiourea sulphate (ZTS) belongs to this class, with zinc bonded to three thiourea molecules and a sulphate ion. One of the advantages of semi organics is that the bonding schemes are three dimensional, unlike polar organic crystals and hence result in stubbier habit. Presently, inorganic and organic materials are replaced by semi organic materials due to their better mechanical and thermal properties along with good second harmonic generation (SHG) nonlinearity.

## 1.8 INTRODUCTION TO COMPUTATIONAL QUANTUM CHEMISTRY

Computational chemistry is a branch of chemistry that uses computer simulation to assist in solving chemical problems. It uses methods of theoretical chemistry, incorporated into efficient computer programs, to calculate the structures and properties of molecules and solids. While computational results normally complement the information obtained by chemical experiments, it can in some cases predict unobserved chemical phenomena. It is widely used in the design of new drugs and materials.

Examples of such properties are structure i.e., the expected positions of the constituent atoms, absolute and relative interaction energies, electronic charge distributions, dipoles and higher multipole moments, vibrational frequencies, reactivity or any other spectroscopic quantities, and cross sections for collision with other particles.

The methods employed cover both static and dynamic situations. In all cases the computer time and other resources (such as memory and disk space) increase rapidly with the size of the system being studied. That system can be a single molecule, a group of molecules, or a solid. Computational chemistry methods range from highly accurate to very approximate; highly accurate methods are typically feasible only for small systems. The modeling approach in material research is greatly facilitated by improvements and accessibility of computer power and emergence of user-friendly molecular modeling packages.

There are three levels in molecular modeling, namely Molecular Mechanics, Semi-empirical and ab initio calculations. Molecular mechanics uses classical physics to explain and interpret the behavior of atoms and molecules. Semi-empirical techniques use approximations from empirical (experimental) data to provide the input into mathematical models. Among these molecular modelling methods, ab initio calculations are most widely applied in energy parameter calculations because of its simplicity and the possibilities of handling large systems in the ground state. Ab initio quantum chemical methods have emerged as an independent and powerful tool for obtaining force constants for any polyatomic molecules. The determination of vibrational frequencies using the ab initio computational method has become increasingly important in many areas of materials physics and chemistry. The various stages of ab initio computational method are, Moller-Plesset (MP), Quadratic configuration interaction (QCISD), Hartree-Fock (HF) and Density functional theory (DFT). In DFT, the total energy is expressed in terms of the total one-electron density rather than the wave function. In this type of calculation, there is an approximate Hamiltonian and an approximate expression for the total electron density. DFT methods can be very accurate for little computational cost. DFT is used for determining the molecular electronic structure, even though many of the most common functionals use parameters derived from empirical data, or from more complex calculations. Some methods combine the density functional exchange with the Hartree–Fock exchange term and are known as hybrid functional methods. The density functional theory (DFT) method could also be utilized to study the vibrational analysis of the polyatomic molecules as they are the base parameters for exploring the further properties of the molecules.

Gaussian 03W (Frisch et. al., 2004) was used for the quantum chemical calculations discussed in this thesis. The Gaussian package contains a myriad of ab initio and semi-empirical methods, although the focus of this thesis is only on density functional theory.

## 1.9 BASIS SET

A basis set in computational chemistry is a set of functions used to create the molecular orbitals, which are expanded as a linear combination of such functions with the weights or coefficients to be determined. In modern computational chemistry, quantum chemical calculations are typically performed within a finite set of basis functions. In these cases, the wave functions under consideration are all represented as vectors, the components of which correspond to coefficients in a linear combination of the basis functions in the basis set used. The operators are then represented as matrices, (rank two tensors), in this finite basis. It should be noted that these basis functions are usually but not actually the exact atomic orbitals, even for the corresponding hydrogen-like atoms, due to approximations and simplifications of their analytic formulas. If the finite basis is expanded towards an infinite complete set of functions, calculations using such a basis set are said to approach the basis set limit.

A basis set can be improved by increasing the number of basis functions per atom. Polarized basis set allow for the addition of orbitals with angular momentum beyond what is required for the ground state description of each atom; this allows for flexibility in different bonding situations. The polarized basis set 6-31G* is a basis set that adds d polarization functions on each nonhydrogen atom. The 6-31G** basis set adds p functions to the hydrogen atom. The 6-31+G** basis set adds diffuse functions (+) to the non-hydrogen atoms, which are important for systems with lone pairs, anions and some excited states, as well as the polarization functions. The 6-311G** basis set is commonly used for electron correlation calculations on molecules containing first-row atoms. The 6-311G** basis set, which is single zeta for the core and triple zeta for the valence atomic orbitals, contains five d-type Gaussian polarization functions on each non-hydrogen atom and three p-type polarization functions on each hydrogen atom

(Pople and Beveridge,1970). Even larger basis sets are available which add multiple polarization functions per atom the triple zeta basis set (Foresman and Frisch, 1993) or additional functions for the valence shell.

## 1.10 DENSITY FUNCTIONAL THEORY (DFT)

Density functional theory (DFT) is a computational quantum mechanical modelling method used in physics, chemistry and materials science to investigate the electronic structure (principally the ground state) of many-body systems, in particular atoms, molecules and the condensed phases. With this theory, the properties of a many-electron system can be determined using functionals, i.e. functions of another function, which in this case is the spatially dependent electron density. Hence the name density functional theory comes from the use of functionals of the electron density. DFT is among the most popular and versatile methods available in condensed-matter physics, computational physics and computational chemistry.

DFT has been very popular for calculations in solid-state physics since the 1970's. However, DFT was not considered accurate enough for calculations in quantum chemistry until the 1990, when the approximations used in the theory were greatly refined to better model the exchange and correlation interactions. In many cases the results of DFT calculations for solid-state systems agree quite satisfactorily with experimental data. Computational costs are relatively low when compared to traditional methods, such as Hartree–Fock theory and its descendants based on the complex many-electron wave function.

Computational methods based on DFT are attractive because they tend to be less resource intensive than conventional ab initio calculations such as Moller-Plesset (MP) and Quadratic configuration interaction (QCISD) while also taking into account the effects of electron correlation. DFT was established by W. Kohn,

P. Hohenberg and L. Sham (Kohn and Sham, 1965) who showed that the electron density could be used as a fundamental quantity to develop a rigorous many-body theory. DFT is different from Hartree-Fock (HF) since it replaces the exchange-correlation energy by a one-electron integral involving the local electron spin densities or by an integral involving these spin densities and their gradients. When these functional are used in conjunction with the Kohn-Sham (KS) procedure (Kohn and Sham, 1965) and an expansion of the KS orbitals in terms of a basis set, then the techniques become analogous to conventional HF theories with the addition of electron correlation (Johnson et. al., 1993).

Formal DFT methods are based on the local density approximation (LDA) given by the following,

$$E_{xc}^{LDA}[\rho(r)] = \int \rho(r)\varepsilon_{xc}(\rho\,(r))dr \qquad (1.23)$$

where $\rho(r)$ the electron density, $\varepsilon_{xc}$ the accurately known exchange-correlation energy per particle of a uniform electron gas of density *n*. The LDA is exact for a homogeneous system and arbitrarily accurate for a system with varying density. Therefore LDA is the most basic approximation for E since it requires only the energy ($\varepsilon_{xc}$) of the uniform electron gas and is found to yield only fair agreement with experimental molecular properties. The local spin-density approximation (LSDA) is an improvement on LDA since it uses different orbitals and different densities for electrons with different spin (Levine, 1991). A better functional would exhibit effects of the inhomogeneity of the electron density, since atoms and molecules do have inhomogeneous electron densities, but this effect is ignored in the LDA model.

The addition of gradient corrections to the LDA model allows for better modeling of the electron density (Parr and Yang, 1989). A gradient-corrected DFT method calculations gives not only the magnitude of the density, but also considers how rapidly the density is changing about a point in space. The more common hybrid DFT methods incorporate some HF exchange with the electron density calculation which includes gradient corrections.

The functionals that are used in DFT are separated into exchange and correlation parts. The exchange part can be either Slater (S), which corresponds to a free-electron gas or Becke (B), which includes a gradient correction. The correlation part can be either ignored or treated with Vosko, Wilk and Nusair (VWN) (Vosko et. al., 1980) parameterization for exact uniform gas results or treated with the gradient corrected functional of Lee, Yang, and Parr (LYP) (Lee et. al., 1988). A common gradient corrected DFT method uses the BLYP functionals which combines a local exchange functional with the gradient correction from Becke and uses the LYP correlation functional. The B3LYP method, the Becke 3 exchange functional coupled with the correlation treatment from LYP, is the DFT method that will be used in this work. The B3LYP method differs from the BLYP method because it utilizes a functional that is a hybrid of exact (HF) exchange with local and gradient-corrected exchange and correlation terms (Stephens et. al., 1994).

DFT methods are appealing for many types of molecules because they can provide results that are as accurate as conventional methods, but DFT is much less resource intensive than such methods as MP, CC (Coupled Cluster) and QCISD (Foresman and Frisch, 1993). Current evidences suggests that vibrational frequencies calculated using DFT is close to those found by experiment.

The computer programs that actually solve the DFT equations have a lot in common with Hartree-Fock programs; quite frequently theorists simply adapt their Hartree-Fock program to do DFT computations. Like Hartree-Fock, the DFT equations must be solved self-consistently, making DFT another type of self consistent field (SCF) method. Instead of focusing on wave functions and orbitals, DFT focuses on the electron density (although it usually employs orbitals to get the density). It includes an approximate treatment of electron correlation and therefore should be more accurate than Hartree-Fock theory. There are actually very many different DFT methods, depending on the particular treatment of correlation or ``exchange". The reason anyone uses DFT is that it tends to give very accurate results much more cheaply than some competing methods.

## 1.11 APPLICATIONS OF GAUSSIAN 03W

Gaussian 03W is one in a series of programs designed, improved and expanded over many years. Gaussian 03W was used exclusively for the calculations in this work and was utilized to visualize optimized geometries and calculated frequencies from Gaussian output files which are the basic parameters required for the interpretation of geometrical molecular structures and computing the nonlinear optical parameters, such as polarizability, dipole moment and hyperpolarizability and other parameters.

### 1.11.1 Geometry Optimization

A geometry optimization can be used to locate minima on a potential energy surface (PES) and can predict the equilibrium structure of the molecule. A point on a PES where the forces are zero is called a stationary point and these are the points generally located during an optimization. Whether these points are local or global minima or even transition states, is another matter and this is represented in figurc 1.3.

An input geometry is provided for geometry optimization and the calculation proceeds to move across the PES. At each point the energy and the gradient are calculated and the distance and direction of the next step are constants that specify the curvature of the surface at the point; this provides additional information useful to determine the next step. Convergence criteria about the forces at a given point and the displacement of the next step determine whether a stationary point has been found (Colthup et. al., 1964).

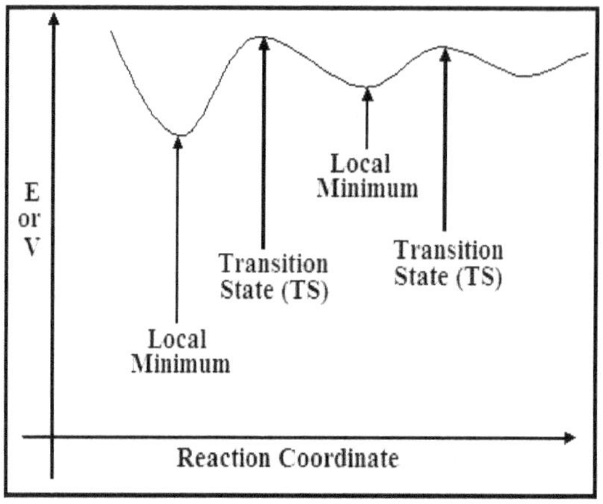


**Figure 1.3 A two-dimensional potential energy curve that shows transition states and local minima**

To determine whether the geometry optimization has found a minimum or a transition state (TS), it is necessary to perform frequency calculations. Transition state is a point that links two minima on the PES (Figure 1.3), and is characterized by one imaginary frequency. The eigen vector from the Hessian force constant matrix determines the nature of the imaginary frequency and indicates a possible reaction coordinate. A minimum structure will have no imaginary frequencies. Distinguishing local minima from global minima relies on chemical intuition and a thorough search of alternative structures.

## 1.12 SCOPE OF THE THESIS

The NLO single crystals of N,N'-diphenylguanidinium nitrate (DPGN), N,N'-diphenylguanidinium hydrogen (+) -L- tartrate monohydrate (DPGTM) and N,N'-diphenylguanidinium dihydrogen phosphite (DPGP) have been studied and identified as potential NLO candidates for device applications. In view of potential scientific interest on NLO crystals and their increasing demand for applications, the present investigation was aimed at the:

(i). Synthesis, growth and study of fundamental parameter, such as solubility and analyzing the morphology of the grown crystals.

(ii). Elucidating the structure of the grown N,N'-diphenylguanidine series crystals through single crystal X-ray diffraction analysis and comparing the structure with those that of the computed ones.

(iii). Analyzing the vibrational structure and hydrogen bonding interactions of the above mentioned compounds through FT-IR and FT-Raman spectroscopic techniques.

(iv). Investigating the optical and nonlinear optical properties of the grown crystals to realize their applications in photonic devices.

(v). Determining the thermal stability of the grown crystals to assess their suitability in device fabrication.

(vi). Understanding the mechanical behaviour of the grown crystals and analyzing the dielectric property of the grown crystals with respect to frequency and temperature.

## 1.13 CONCLUSION

A brief introduction on crystal growth and classification of crystal growth methods have been described. Various methods of crystal growth, especially the fundamental and experimental aspects of low temperature solution growth technique have been explained in detail. The fundamental principle and applications of nonlinear optical phenomena and an insight into different NLO materials have been discussed. The basic outline on computational chemistry and density functional theory have been discussed.

# CHAPTER – II

# SYNTHESIS AND GROWTH OF N,N'-DIPHENYLGUANIDINE BASED NLO SINGLE CRYSTALS

## 2.1 INTRODUCTION

Nonlinear optical (NLO) materials, which can generate second harmonic blue-violet light, are of great interest for various applications including optical communication, optical computing, optical information processing, optical disk data storage, laser fusion reactions, laser remote sensing, color display, medical diagnostics etc.

Guanidinium family complex has been chosen for study as potential material for non-linear optics (NLO). Guanidine can be employed in material engineering of a promising class of NLO compounds. They are based on salts combining a cation derived from a polarizable organic molecule with an anion capable of forming hydrogen-bonded crystal structures (Aakeroy et. al., 1993a). The ability of guanidinium ion, in making hydrogen bonds and its unique planar shape has been recognized by Drozd et. al., (2005a). The crystals of guanidinium compounds, display interesting physical and chemical properties, exhibiting phase transitions with ferroelectric, anti-ferroelectric and ferro-elastic materials. From the point of view of their physical properties, guanidine compounds are potentially interesting for non-linear applications. Non-linear materials are those in which the dielectric polarization responds nonlinearly to the electric field of the incident light, producing different frequencies. Such materials are used to associate with lasers enabling the expansion of the lasers limited spectral regime.

Guanidine is also an important compound that has many biological, chemical and medical applications. Guanidinium derivatives are of great biological importance due to its presence as the functional group of amino acid arginine, the muscular energy-intermediate creatine, the pyrimidine bases of DNA and a number of other biologically active molecules (James et. al., 2007) and because of its use as a protein denaturant. Potential guanidine family candidates exhibiting nonlinear optical property, such as guanidinium 4-nitrobenzoate (Arumanayagam and Murugakoothan, 2013), bis guanidinium hydrogen phosphate monohydrate (Russel and Murugakoothan, 2013) and guanidinium 3-nitrobenzoate (Suvitha et. al., 2013) were reported earlier.

Salts of guanidine and guanidine derivatives have shown interesting NLO properties that prompted the interest in probing the salts of guanidine derivative i.e., N,N'-diphenylguanidine also known as melaniline that has been widely used as a primary and secondary accelerator in the vulcanization of rubber. It also acts as a catalyst in the synthesis of sulfur-containing compounds, such as thiols, thiazoles, sulfonamides and thiurams. Development of many new compounds from guanidine cation and its derivatives has been recently emerging on a wide scale. In the present work, series of three compounds with N,N'-diphenylguanidine cations were developed and investigated as potential materials for nonlinear optics. This chapter deals with the synthesis and growth of N,N'-diphenylguanidine based single crystals by low temperature solution method using slow evaporation technique.

Growth of crystals from aqueous solution is one of the ancient methods of crystal growth. The method of crystal growth from low temperature aqueous solution is an universal method which is extremely popular in the production of many technologically important crystals. It is the most widely used method for the growth of single crystals, when the starting materials are unstable at high temperatures and undergo phase transformations below the melting point. Materials having moderate to high solubility in temperature range, ambient to 100 °C at atmospheric pressure, can be grown by low-temperature solution method. Since growth is carried out at near room temperature, the structural imperfections in solution grown crystals are relatively low (Hooper et. al., 1980; Brice, 1972). Detailed studies of the crystals can provide clues to adopt the required modifications in growth techniques in order to improve the perfection of the crystals. It is also important to know the nature and distribution of the imperfections present in the crystals. A thorough assessment only can determine the quality and perfection of the crystals.

## 2.2 MATERIAL SYNTHESIS

In order to achieve good nonlinear response in organic crystals, one requires the presence of delocalized π-electron systems connecting donor and acceptor groups. This system enhances the necessary molecular hyperpolarizability and also proper orientation of the molecules in the solid state structure to facilitate high frequency conversion efficiency.

N,N'-diphenylguanidine or DPG is an organic compound with the molecular formula $C_{13}H_{13}N_3$. The substance occurs as a colorless to white crystalline powder that is insoluble in water. It is however, quite soluble in ethanol, acetone, chloroform and carbon tetrachloride . N,N'-diphenylguanidine is a strong base and readily react with all types of organic and inorganic acids largely because of the presence of potential donor sites for hydrogen bonding interactions. The short C-N bonds exhibit a pronounced electron-acceptor π-bond character. Hence, N,N'-diphenylguanidine cation is protonated with some inorganic and organic acids, such as nitric acid, phosphorous acid and L-tartaric acid, to yield good quality crystals namely,

(i). N,N'-diphenylguanidinium nitrate (DPGN).

(ii). N,N'-diphenylguanidinium dihydrogen phosphite (DPGP) and

(iii). N,N'-diphenylguanidinium hydrogen (+) -L-tartrate monohydrate (DPGTM).

All the above mentioned compounds were synthesized by slow evaporation technique with a stoichiometric (1:1) incorporation of N,N'-diphenylguanidine with their respective acids in different solvents at room temperature. The basic information of the DPGN, DPGP and DPGTM are provided in Table 2.1.

**Table 2.1 Basic information of title compounds**

| DATA | N,N'-diphenylguanidinium nitrate (**DPGN**) | N,N'-diphenylguanidinium dihydrogen phosphite (**DPGP**) | N,N'-diphenylguanidinium hydrogen(+)-L-tartrate monohydrate (**DPGTM**) |
|---|---|---|---|
| Base | N,N'-diphenylguanidine | N,N'-diphenylguanidine | N,N'-diphenylguanidine |
| Acid | Nitric acid | Phosphorous acid | L-Tartaric acid |
| Solvent | Methanol | Water + Ethanol (1:1) | Water + Ethanol (1:1) |
| Molecular formula | $C_{13}H_{14}N_3^+.NO_3^-$ | $C_{13}H_{14}N_3^+.\ H_2PO_3^-$ | $C_{13}H_{14}N_3^+.C_4H_5O_6.H_2O$ |
| Molecular Weight | 274.28 | 293.26 | 379.37 |
| Molecular structure | 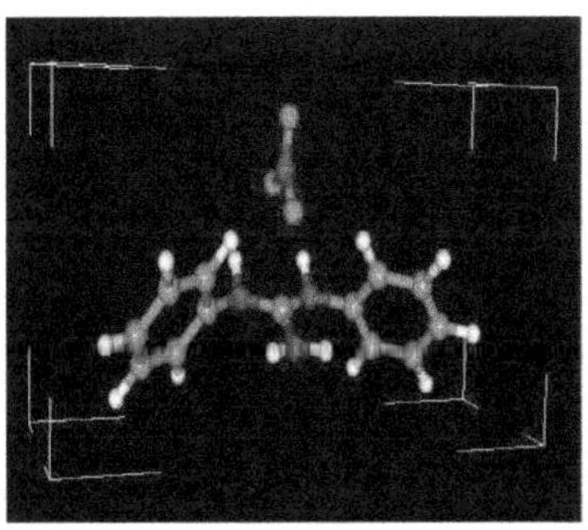 | 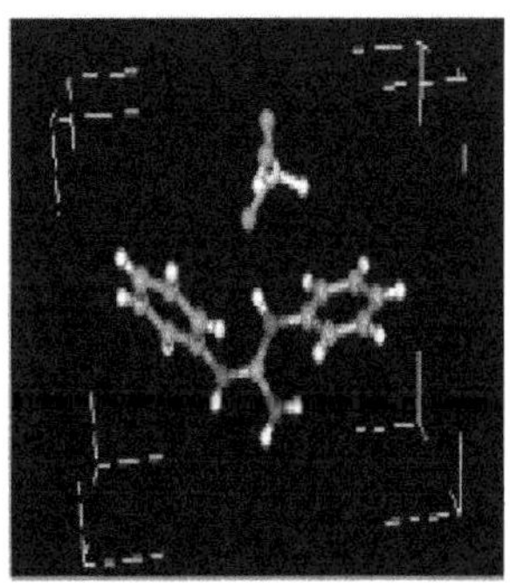 | 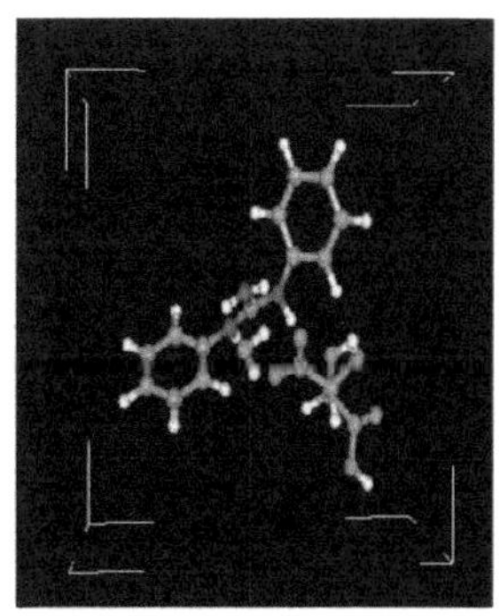 |

In the synthesis of DPGN the starting materials were N,N'-diphenylguanidine and nitric acid. The calculated amount of nitric acid was first dissolved in methanol and was made to stir continuously at room temperature. With constant stirring, the calculated amount of N, N'-diphenylguanidine was added to the methanolic solution of nitric acid resulting in precipitate formation. The formation of precipitate confirms the compound formation and the precipitate was made to dissolve by adding excess of methanol until complete neutralization occurred thereby resulting in pale brown colored solution. The chemical reaction is given below.

$$C_{13}H_{13}N_3 + HNO_3 \longrightarrow C_{13}H_{14}N_3^+ \;.\; NO_3^-$$

N,N'- Diphenylguanidine + Nitric acid ⟶ N, N'- Diphenylguanidinium nitrate

The compounds of DPGTM and DPGP were synthesized by reacting stoichiometric parts of N,N'-diphenylguanidine with L – tartaric acid and phosphorous acid respectively, from a mixed solvent of water + ethanol (1:1). The calculated amounts of respective acids were first dissolved in water. To this aqueous solution of acids the calculated amount of ethanolic solution of N,N'-diphenylguanidine was added until complete neutralization takes place. The precipitate formed as a result of neutralization thereby indicating the compound formation. The precipitate was dissolved by adding excess of mixed solvent at room temperature. The respective chemical reactions are given below.

$$C_{13}H_{13}N_3 + C_4H_6O_6 \xrightarrow{\text{water}} C_{13}H_{14}N_3^+.\; C_4H_5O_6^-.\; H_2O$$

N,N'- Diphenylguanidine + L-tartaric acid $\xrightarrow{\text{water}}$ N,N'- Diphenylguanidinium hydrogen L-tartrate monohydrate

$$C_{13}H_{13}N_3 + H_3PO_3 \longrightarrow C_{13}H_{14}N_3^+.\; H_2PO_3^-$$

N,N'- Diphenylguanidine + Phosphorous acid ⟶ N,N'- Diphenylguanidinium dihydrogen phosphite

## 2.3 SOLUBILITY STUDIES OF N,N'–DIPHENYLGUANIDINE BASED CRYSTALS

Solubility is an important parameter, which converses the growth procedure. If the solubility is too high, it is difficult to grow bulk single crystals with optical quality and too low solubility restricts the size and growth rate of the crystals. Neither a flat nor a steep solubility curve will enable the growth of bulk crystals from solution.

The low temperature solution growth is mainly a diffusion-controlled process, in which the medium must be less viscous to enable faster transfer of the growth units through the bulk solution by diffusion. Hence a solvent with less viscosity is preferable. Solubility corresponds to saturation i.e., equilibrium between a solid and its solution at a given temperature. Thermodynamically, saturation means the chemical potential of the pure solid is equal to the chemical potential of the same solute in the saturated solution. Solubility data of a material governs the amount of material, which is available for the growth and hence defines the total size limit. Solvent and solubility factors define the supersaturation, which is the driving force for the rate of crystal growth. Hence for a material to be grown as a crystal, determination of its solubility in a particular solvent is an essential criterion (Christian, 1990).

The solubility test of DPGN was carried out with methanol as solvent. The prepared solution of DPGN was continuously stirred at a constant temperature of 30 °C using a motorized magnetic stirrer to ensure homogeneous temperature and concentration over the entire volume of the solution. The saturated solution was kept at a particular chosen temperature for about 6 hours to attain equilibrium. A 10 mL of solution was taken in a warm pipette and poured in to petridish. The solubility was gravimetrically (Nyvlt et. al., 1970) analyzed by evaporating the solution in an oven at a constant temperature. The same procedure was repeated for different temperatures, such as 35, 40, 45 and 50 °C. The same procedure was adopted for the determination of solubility of DPGTM and DPGP in a mixed solvent of water + ethanol (1:1). It is observed that the solubility increases linearly with temperature for all the compounds studied in this work. Higher the solubility of a material higher the growth rate of a crystal at an optimized condition. The moderate solubility of the compounds indicates the reasonable growth rate along all crystallographic directions. The solubility of N,N'- diphenylguanidine based compounds for various solvents are depicted in figure 2.1.

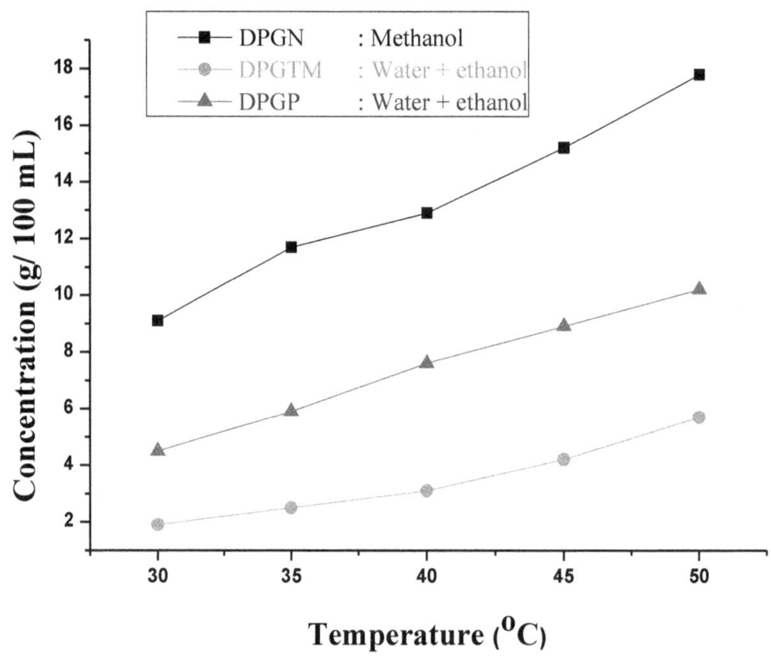


**Figure 2.1 Solubility curve of DPGN, DPGTM and DPGP**

It is clear that DPGN has higher solubility at 35 °C when compared with other N,N'-diphenylguanidine compounds as infered from figure 2.1. The solubility data for above said compounds at lower temperature range is depicted in Table 2.2.

**Table 2.2 Solubiltiy data of title compounds at 35 °C**

| S.No. | COMPOUND | SOLVENT | SOLUBILITY (g/100 mL) |
|---|---|---|---|
| 1. | N,N'-diphenylguanidinium nitrate | Methanol | 11.7 |
| 2. | N,N'-diphenylguanidinium hydrogen (+)-L-tartrate monohydrate | Water + ethanol (1:1) | 2.5 |
| 3. | N,N'-diphenylguanidinium dihydrogen phosphite | Water + ethanol (1:1) | 5.9 |

## 2.4 CRYSTAL GROWTH

The slow evaporation technique was adopted for this work to grow crystals. As this techinque is found to be the best as the crystals are allowed to grow in natural manner and ambient environmental condition without any restrictions. The solutions of the title compounds were prepared by dissolving the respective salts, such as recrystallized DPGN in methanol and DPGTM and DPGP in mixed solvent of water+ethanol (1:1). The solutions were stirred for 6 hours to achieve homogenization in both temperature and concentration throughout the solution, using magnetic stirrer at room temperature. The corresponding solutions were filtered in a clean beaker through whatmann filter paper. The beaker was covered with aluminium foil in order to control the evaporation rate. The optically transparent, non-hygroscopic and good quality single crystals of DPGN, DPGTM and DPGP were grown by slow evaporation method. The DPGN, DPGTM and DPGP crystals were harvested after a period of 46, 90 and 36 days respectively. The as grown single crystals of DPGN, DPGTM and DPGP are shown in figure 2.2.

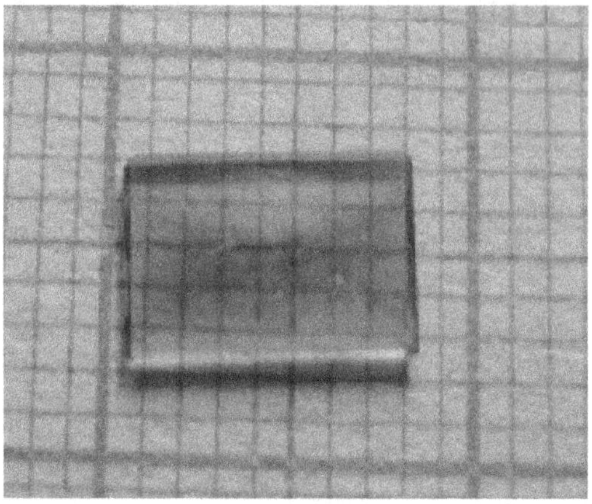

DPGN

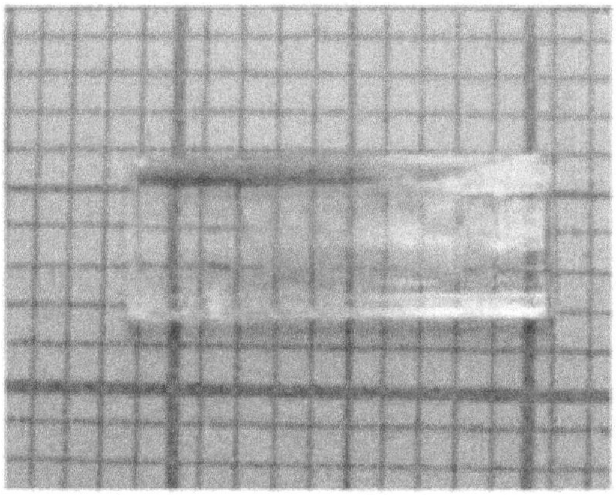

DPGTM

DPGP

**Figure 2.2 Single crystals of DPGN, DPGTM and DPGP grown by slow evaporation tecchnique**

## 2.5 MORPHOLOGY

The studies on surface morphology of crystals are useful to know about the mechanism of crystal growth and thereby to eliminate the defects, enhance the growth rate and improve the quality (Jiang et. al., 2002). The morphology of growing crystal is broadly governed by two factors, the internal structure of the crystal and external parameters, such as supersaturation, temperature level, orientation of the seed, the presence of solvent and impurities. The morphology of a crystal is the result of the relative growth rates of its various facets, and the general rule being that facets which grow slowest are expressed in the crystal habit. The growth rates of the different crystal facets are determined by intermolecular interactions between molecules in the crystal, as well as by a number of experimental parameters, such as solvent supersaturation, pH value, temperature, impurities and hydrodynamics (Zhang and Zhang, 1997). The important parameters controlling the growth rate are the energy required to create a step at the crystal surface and free energy barrier for an adsorbed impurity ions to be incorporated in to the crystal. If the surface relaxations are large enough then small nuclei will grow into platelets, which further grow into large three dimensional crystals (Lee et. al., 1999).

The morphology and crystallographic faces of the title crystals were studied by single crystal X-ray diffraction using ENRAF NONIUS CAD4 automatic X-ray diffractometer with MoK$\alpha$ ($\lambda$ = 0.7170 Å) radiation. The morphology of as grown single crystal of DPGN is shown in figure 2.3(a). The morphology of the crystallographic faces of the grown DPGN crystal exhibits seven well developed faces. Among the well-developed faces in DPGN the (010) face is larger in area. The morphology of DPGN crystal indicates that the growth rate along the 'a' direction is faster than that on the other directions, hence DPGN crystal is elongated along the crystallographic a-axis.

The morphology of DPGTM crystal, shown in figure 2.3(b), is a prism with five well-developed faces. The morphology of a crystal is the result of the relative growth rates of its various facets, and facets which grow fast, slowly extinguishes with growth rate. Therefore, it also indicates that the growth along the a-axis of the crystal is faster than that of the other two crystallographic axis.

The morphology of DPGP, shown in figure 2.3(c) establishes that there are twelve developed faces, out of which (110) face is more prominent. For each face, its parallel friedal plane is also present in the grown crystal.

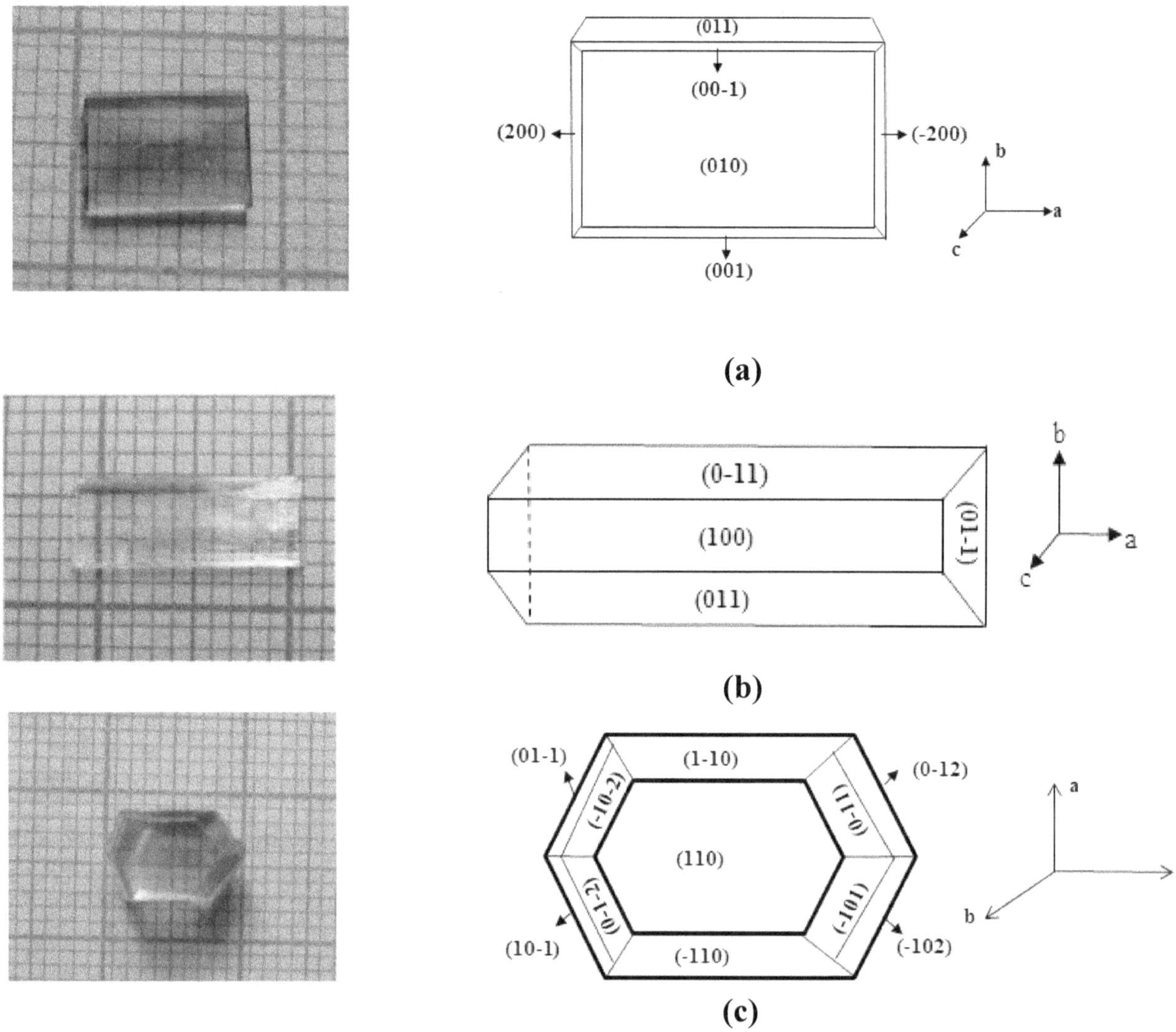


**Figure 2.3 Morphology of as grown (a) DPGN crystal, (b) DPGTM crystal and (c) DPGP crystal**

## 2.6 CONCLUSION

The N,N'-diphenylguanidine based compounds, such as DPGN, DPGTM and DPGP were synthesized by low temperature solution growth using slow evaporation technique. The solubility of title compounds was determined for their corresponding solvents at different temperatures and it indicates that the solubility increases linearly with temperature. The synthesized salts were used to grow the single crystals of the title compounds. The grown crystals are observed to be non-hygroscopic and transparent with well-defined morphology. The morphological analysis was carried out to assess the facets of the grown crystals.

# CHAPTER – III
# STRUCTURAL AND SPECTRAL ANALYSES OF N,N'-DIPHENYLGUANIDINE BASED NLO SINGLE CRYSTALS

## 3.1 INTRODUCTION

Design engineers, Scientists and Researchers use a wide range of instrumentation technique to assess the suitability of the grown crystals for employing them in device fabrication. Right from the synthesis to device fabrication of a crystal, various characterization technique are involved in order to predict their standards in each and every step of the processes. The purity, mechanical stability, thermal stability and other necessary parameters of the crystals are very essential to employ them successfully in devices. The processing of the crystal in device applications is the only fruitful achievement for the purpose of crystal growth.

Characterization is the assessment technique to reveal the physical and chemical properties of a material both qualitatively and quantitatively. The assessment technique assists us to trigger rapid progress in the growth and to improve the quality of the crystals. According to Elwell and Scheel (1975), characterization of crystals essentially consists of an evaluation of its chemical composition, structure, defects, optical properties and other key parameters.

As guanidine and their family members are subject of interest in the field of nonlinear optics, N,N'-diphenylguanidine is a family member of guanidinium cation with two phenyl rings attached to it. The family of compounds based on the cations of guanidinium and its derivatives are the subject of chemical research for their role in living organisms, molecular design and as a potential denaturant of proteins and nucleic acids (Von Hippel and Schleich, 1969; Makhatadze, 1999). Hydrogen bonding in these salts provides a notable energetic contribution to the total lattice energy (Aakeroy and Seddon, 1993b) and contributes to the second-order NLO tensor coefficient ($d_{ijk}$) of the crystals. The family of compounds with guanidinium cation belongs to the simplest organic chemical compounds, play a vital role in the features of investigated crystals suitable for nonlinear optics (NLO) (Drozd and Baran, 2005b). The guanidinium ion is characterized by a relatively small ionic radius. On the other hand, the specific planar configuration makes the guanidinium cation as potential H-donor in hydrogen bonds (Nalwa et. al., 1997b). These properties can be employed in molecular engineering where many new compounds with programmable features are synthesized.

In this chapter, our attention is to throw light on the structural and spectral investigations on N,N'-diphenylguanidine based compounds. The structural investigation comprises of single crystal X-ray diffraction (SXRD) and powder X-ray diffraction (PXRD) analyses which were employed to study the structural parameters and crystalline nature of the title compounds. The Density Functional Theory (DFT) was employed using Gaussian 03W computational software to simulate the optimized molecular structure of the title compounds in order to correlate the stimulated structure with that of the experimental ones as the optimized molecular structure forms the base for all further theoretical studies. From the spectral analyses, we can obtain more accurate information on its structure and obtain evidence of the molecular interactions and vibrations of the compounds. The spectral studies taken up for the analysis of N,N'-diphenylguanidine series crystals are, namely Fourier transform infrared spectroscopy and Fourier transform Raman spectroscopy.

## 3.2 SINGLE CRYSTAL X-RAY DIFFRACTION ANALYSIS

### 3.2.1 Introduction

The structure of a crystal is the basis for understanding many of its properties. Single crystal X-ray diffraction is a non-destructive analytical technique, which provides detailed information about the internal lattice of crystalline substance, including unit cell dimensions, bond-lengths, bond-angles and details of site ordering. The data generated from the X-ray analysis is interpreted and refined to obtain the crystal structure.

Max Von Laue, in 1912, discovered that crystalline substance act as three-dimensional diffraction grating for X-ray wavelengths. X-ray diffraction is now a common technique for the study of crystal structure and perfection of atomic packing. X-ray diffraction is based on constructive interference of monochromatic X-rays diffracted by a crystalline sample. X-rays are generated by a cathode ray tube, filtered to produce monochromatic radiation, collimated to concentrate and directed toward the sample. The interaction of the incident rays with the sample produces constructive interference when conditions satisfy Bragg's Law

$$2d \sin\theta = n\lambda \qquad (3.1)$$

This law relates the wavelength of electromagnetic radiation to the diffraction angle and the lattice spacing in a crystalline sample. These diffracted X-rays are then detected, processed and counted. By changing the geometry of the incident rays, the orientation of the centered crystal and the detector, all possible diffraction planes of the crystal can be obtained. Typical crystal structures contain several thousand unique reflections, whose spatial arrangement is referred to as a diffraction pattern. Indices (h k l) may be assigned to each reflection, indicating its position within the diffraction pattern. After the structure is solved, it is further refined using least-squares techniques.

There are three main X-ray diffraction methods by which the crystal structures can be analyzed, they are :

(i). the Laue method which is applicable to single crystals,

(ii). the rotating crystal method which is applicable to single crystals and

(iii). the powder method which is applicable to finely divided crystalline or polycrystalline specimen powder.

Single crystal X-ray diffraction technique is most commonly used for precise determination of a unit cell including cell dimensions and position of atoms within the lattice. Bond-length and bond angles are directly related to the atomic positions. Specific applications of single crystal X-ray diffraction include:

(i). new materials identification,

(ii). precise determination of the three dimensional crystal structure of a compound,

(iii). determination of unit cell, bond-lengths, bond-angles and site ordering,

(iv). characterization of cation-anion coordination,

(v). magneto-structural correlations in molecular materials and

(vi). absolute configuration determination.

### 3.2.2 Rotating Crystal Method

The rotating crystal method can be applied only if single crystals of reasonable size are available. In this method, a single crystal is rotated about the fixed axis in a beam of monochromatic X-rays. The variation in θ brings different atomic planes into position for reflection. The reflections are usually recorded on cylindrical film whose axis is the crystal. The reflected spots lie on parallel lines. Since the wavelength is known, inter planar spacing may be calculated using Bragg's law. The rotation method furnishes the shape and size of the unit cell of unknown crystal provided that the crystal can be oriented to rotate about the requisite.

### 3.2.3 Single Crystal X-ray Diffraction Analysis of N,N'-Diphenylguanidine Based NLO Single Crystals

The single crystal X-ray diffraction analysis of the grown N,N'-Diphenylguanidine series of crystals was carried out using ENRAF NONIUS CAD4 automatic X-ray diffractometer with MoKα (λ = 0.7170 Å) radiation. The unit cell dimensions and the space group of the grown crystals were determined. The DPGN and DPGTM belongs to orthorhombic crystal system with $Pna2_1$ and $P2_12_12_1$ as space group. The DPGP belongs to tetragonal crystal system with $P4_3$ as space group. The observed structural data of DPGN, DPGTM and DPGP are given in Table 3.1. The data obtained from the single crystal XRD study are found to be in good agreement with the literature values (Paixao et. al., 1998, Paixao et. al., 1999, Paixao et. al., 2001). As non-centrosymmetric space group is one of the major requirement for the existence of second harmonic generation activity, all the title compounds found to crystallize in the non-centrosymmetric space group thereby satisfying the requirement.

### 3.2.4 Optimized Molecular Geometry of N,N'-Diphenylguanidine Based NLO Single Crystals

***Computational Details***

For studying the quantum chemical properties of any molecule the first step required is the optimization of the molecular structure. From the optimized molecular structure the other parameters have to be studied and to be interpreted. The DFT calculations were performed on a Pentium IV personal computer using the Gaussian-03W program package (Frisch et. al., 2004). The molecular structure of DPGN, DPGTM and DPGP was optimized using B3LYP, hybrid function consisting of Becke's three-parameter non-local exchange function with the correlation function of Lee, Yang and Parr (Lee et. al., 1988).

**Table 3.1 Single crystal XRD data for the grown crystals**

| Crystal | Unit Cell Parameters | Space Group |
|---|---|---|
| DPGN | a = 16.97 (5) Å<br>b = 13.89 (4) Å<br>c = 5.79 (16) Å<br>V = 1364 (9) $Å^3$<br>α = β = γ = 90° | Orthorhombic<br>*$Pna2_1$*<br>(Paixao et. al., 1998). |
| DPGTM | a = 7.05 (9) Å<br>b = 14.72(19) Å<br>c = 18.20 (2) Å<br>V = 1889 (6) $Å^3$<br>α = β = γ = 90° | Orthorhombic<br>*$P2_12_12_1$*<br>(Paixao et. al., 1999). |
| DPGP | a = 10.82 (2) Å<br>b = 10.82 (2) Å<br>c = 12.64 (3) Å<br>V = 1481 (5) $Å^3$<br>α = β = γ = 90° | Tetragonal<br>*$P4_3$*<br>(Paixao et. al., 2001). |

However for optimizing and obtaining the quantum chemical parameters for DPGTM the same level of B3LYP, hybrid function consisting of Becke's three-parameter non-local exchange function with the correlation function of Lee, Yang and Parr was employed with B3LYP/6-31G (d) basis set.

***Geometry Optimization***

Geometry optimization of DPGN, DPGTM and DPGP was carried out from the X-ray experimental atomic position Crystallographic Information File (CIF) (Paixao et. al., 1988, 1999, 2001) of the corresponding structure and then all other calculations engaging density functional theory (DFT) were performed by the resultant optimized structural parameters. The optimized molecular geometry represents an isolated single molecule under ideal conditions with a stationary point at the potential energy surface; the convergence was confirmed by observing no imaginary vibrational wavenumbers obtained in the calculation.

The self-consistent field (SCF) energy of DPGN and DPGP at B3LYP level with the basis set 6-31 G (d,p) was found to be -948.417 a.u. and -1236.407 a.u. However, the self-consistent field (SCF) energy of DPGTM at B3LYP/6-31 G (d) level was found to be -1351.305 a.u. The optimized molecular geometry of DPGN, DPGTM and DPGP is shown in figures 3.1 – 3.3.

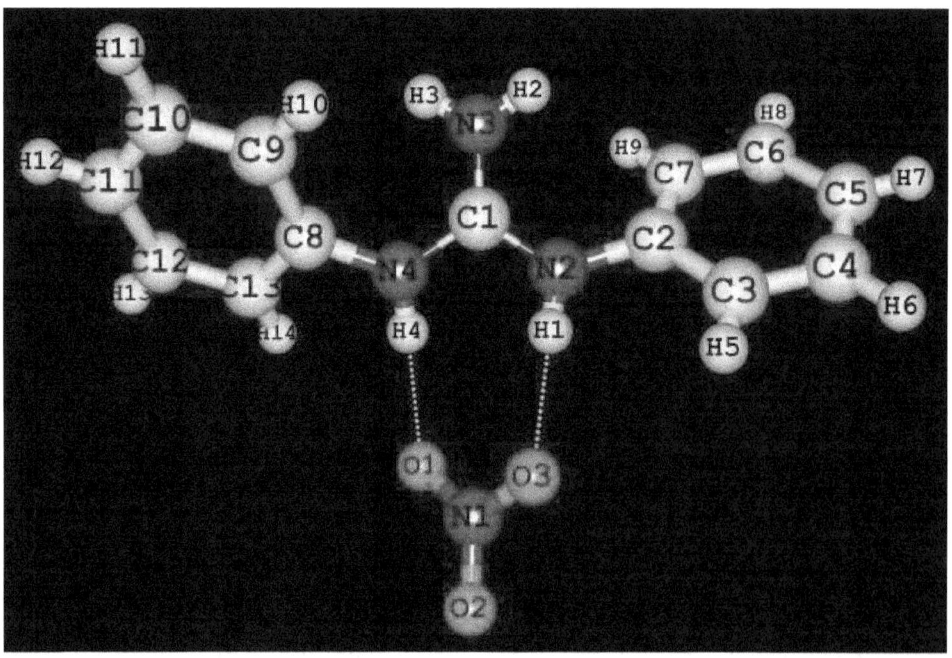


**Figure 3.1 Optimized molecular geometry of DPGN**

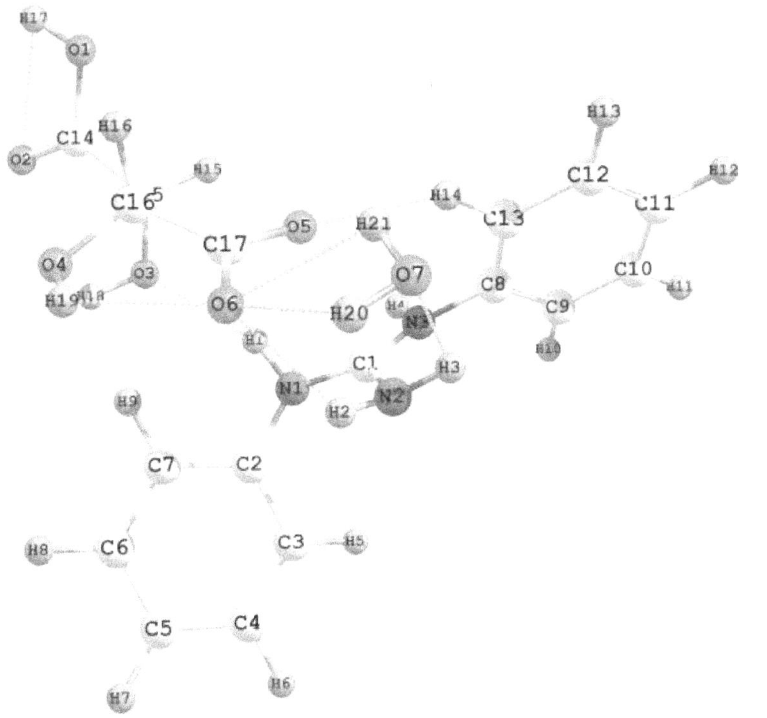


**Figure 3.2 Optimized molecular geometry of DPGTM**

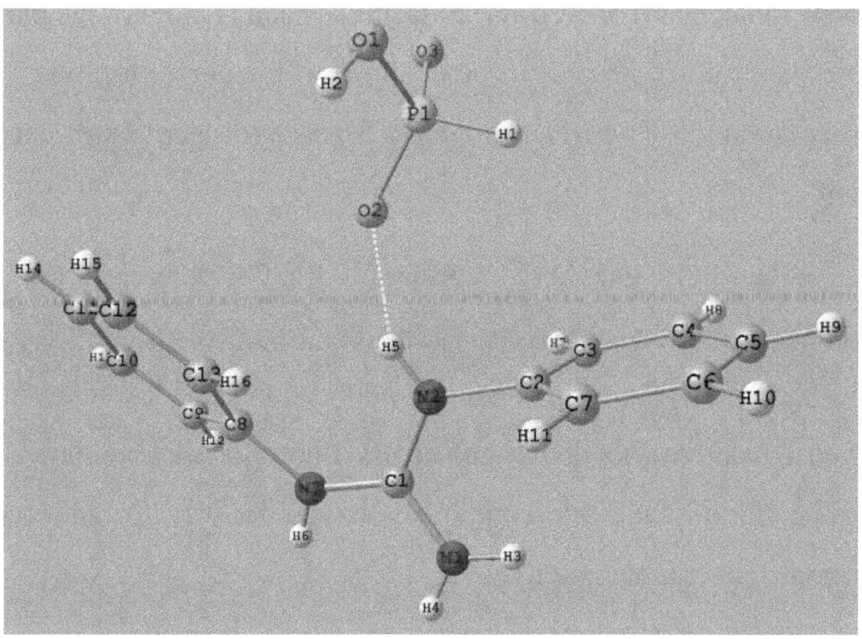


**Figure 3.3 Optimized molecular geometry of DPGP**

The central guanidinium cation fragment of DPGN, is planar and the geometry is close to that expected for a central $C_{sp}^2$, atom. The bond lengths C1 – N2, C1 – N3, C1 – N4 within the range expected for a delocalized bond. With regard to the geometry of the anion, there is a slight dissymmetry in the N – O bond lengths; the N1– O2 bond is significantly shorter than the other two. This probably reflects the fact that the O2 atom is involved in a weaker hydrogen bond than the other O atoms. The nitrate anion and guanidinium cation are linked by a two dimensional network of hydrogen bonds. Each of the N – H groups is involved in a hydrogen bond with oxygen atoms of the anion.

The geometry of the guanidinium group in DPGTM is close to that expected for a central $C_{sp}^2$ atom. The N1-C1 and N3-C1 bond lengths are close to the expected value for a delocalized bond. The anions and cations in DPGTM are held together by an intricate network of hydrogen bonds. The full potential for hydrogen bonding of both the anion and cation is realized in this structure. The strongest hydrogen bond links the carboxyl and carboxylate groups of two neighboring hydrogen tartrate molecules in a head-to-tail fashion. Each N atom of the guanidinium moiety shares its proton with one of the O atoms of the anion, forming moderately strong hydrogen bonds. The water molecule forms hydrogen bonds exclusively with the anion.

The central guanidinium group of DPGP is planar and the C-N bond lengths of the guanidinium group are within the expected range which is typical for a delocalized C-N bond. The anions and cations are linked by an intricate three dimensional network of hydrogen bonds. This hydrogen bonding between the cation and the anion of the title compounds paves the way for enhancement of second harmonic generation activity.

The calculated geometrical parameters and the corresponding experimental values taken from X-ray diffraction data for DPGN, DPGTM and DPGP are listed in Table 3.2 – 3.4. As seen, the optimized geometrical parameters represent good approximation and can be used as a foundation to calculate the other parameters and properties of the title compounds. The slight deviations in theoretical parameters of optimized geometry from the experimental values are probably due to the intermolecular interactions in the crystalline state.

**Table 3.2 Selected optimized molecular geometrical parameters for DPGN molecule**

| Parameters | B3LYP | Experimental (Paixao et. al., 1998) |
|---|---|---|
| Bond | Bond Length (Å) | Bond Length (Å) |
| O1 – N1 | 1.281 | 1.252 |
| O2 – N1 | 1.224 | 1.224 |
| O3 – N1 | 1.280 | 1.260 |
| N2 – C2 | 1.416 | 1.417 |
| N3 – C1 | 1.349 | 1.310 |
| N4 – C1 | 1.340 | 1.348 |
| N2 – C1 | 1.340 | 1.328 |
| N4 – C8 | 1.416 | 1.416 |
| Bond | Bond Angle ($^{O}$) | Bond Angle ($^{O}$) |
| C1–N2–C2 | 127.22 | 130.3 |
| C1–N4–C8 | 127.01 | 128.2 |
| O2–N1–O1 | 120.79 | 119.9 |
| O1–N4–O3 | 120.78 | 119.5 |

**Table 3.3 Selected optimized molecular geometrical parameters for DPGTM molecule**

| Parameters | B3LYP | Experimental (Paixao et. al., 1999) |
|---|---|---|
| Bond | Bond Length (Å) | Bond Length (Å) |
| N1 – C1 | 1.348 | 1.330 |
| N1 – C2 | 1.437 | 1.418 |
| N2 – C1 | 1.335 | 1.314 |
| N3 – C1 | 1.356 | 1.332 |
| N3 – C8 | 1.423 | 1.419 |
| C14 – O2 | 1.214 | 1.213 |
| C14 – O1 | 1.347 | 1.289 |
| C17 – O6 | 1.247 | 1.230 |
| C17 – O5 | 1.278 | 1.276 |
| Bond | Bond Angle ($^{O}$) | Bond Angle ($^{O}$) |
| N2–C1–N1 | 121.3 | 120.6 |
| N2–C1–N3 | 121.2 | 121.8 |

**Table 3.4 Selected optimized molecular geometrical parameters for DPGP molecule**

| Parameters | B3LYP | Experimental (Paixao et.al., 2001) |
|---|---|---|
| Bond | Bond Length (Å) | Bond Length (Å) |
| P1 – O1 | 1.625 | 1.535 |
| P1 – O2 | 1.595 | 1.486 |
| P1 – O3 | 1.491 | 1.487 |
| O1 – H2 | 0.971 | 0.820 |
| N1 – C1 | 1.388 | 1.327 |
| C1 – N2 | 1.295 | 1.324 |
| Bond | Bond Angle ($^{O}$) | Bond Angle ($^{O}$) |
| O1–P1–O2 | 114.3 | 113.6 |
| O1–P1–O3 | 109.2 | 108.0 |
| N3–C1–N1 | 122.2 | 121.4 |

## 3.3 POWDER X-RAY DIFFRACTION

### 3.3.1 Introduction

Crystallinity of the chosen material can easily be determined from powder X-ray diffraction analysis. This powder method is applicable only to finely divided crystalline powder or to a very fine-grained polycrystalline specimen. This method is also called as Debye, Scherer and Hull method. This method is more convenient since single crystals are not required and it is used for accurate determination of lattice parameters in crystal of known structure and for the identification of elements and compounds. The powdered specimen was kept inside a small capillary tube, which does not undergo diffraction. A narrow pencil of monochromatic X-rays is diffracted from the powder and recorded by the photographic film as a series of lines of varying curvature. The full opening angle of the diffraction cone 4θ is determined by measuring the distance 'S' between the two corresponding arcs on the powder photograph symmetrically displaced about the exit point of the direct beam.

The distance 'S' on the film between two diffraction lines corresponding to a particular plane is related to the Bragg angle by the equation.

$$4\theta = \frac{S}{R} \text{ radian} \qquad (3.2)$$

$$4\theta = \frac{S}{R}\left(\frac{180}{\pi}\right) \text{ degree} \qquad (3.3)$$

where R the distance between the specimen and film. A list of θ values can be obtained from the measured values of S. Since wavelength is known, substitution of θ gives a list of spacing d. Each space is the distance of neighboring planes (h k l) (William, 2004; Warren, 1990).

Crystallinity of the chosen material can easily be determined from powder X-ray diffraction analysis. The powder XRD pattern is a plot of peak intensity versus diffraction angle (2θ). The crystalline materials show sharp intense peaks whereas the amorphous salts show only undefined broad peaks of relatively low intensity. The powder XRD pattern exhibits one peak for all repeating planes with equal 'd' spacing irrespective of their orientation. Powder XRD patterns are unique for a given crystal structure and are used to detect changes in the hydrate due to hydration, dehydration or polymorphic transitions. This is accomplished by peak matching, which may be done either manually or by specially designed soft wares.

### 3.3.2 Powder X-Ray Diffraction Study of N,N'-Diphenylguanidine Based NLO Single Crystals

The crystalline quality of N,N'-diphenylguanidine based crystals, grown in the present work, was confirmed through the powder X-ray diffraction (PXRD) analysis. Finely crushed powder of N,N'-diphenylguanidine crystals was subjected to PXRD study using Reich-Seifert diffractometer with CuKα (λ = 1.5418 Å) radiation in the range to 10° – 70° at the scan rate of 2° /minute. The lattice parameter values obtained from the single crystal XRD analysis were used to index the peaks in the powder XRD spectrum using POWDER X program and lattice parameters were evaluated. The sharp peaks found in the spectra shows good crystallinity of the grown crystals. The presence of prominent Bragg peaks at specific 2θ angles in the powder X-ray diffraction pictogram confirms the perfect crystalline nature of the title crystals. The indexed PXRD pattern of the DPGN, DPGTM and DPGP crystalline sample is shown in figures 3.4 – 3.6.

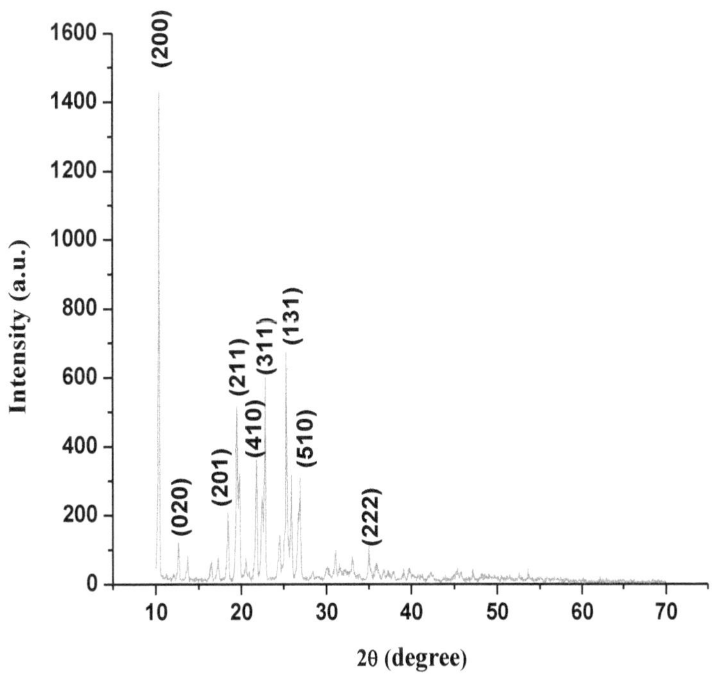


**Figure 3.4 Powder X-ray diffractogram of DPGN**

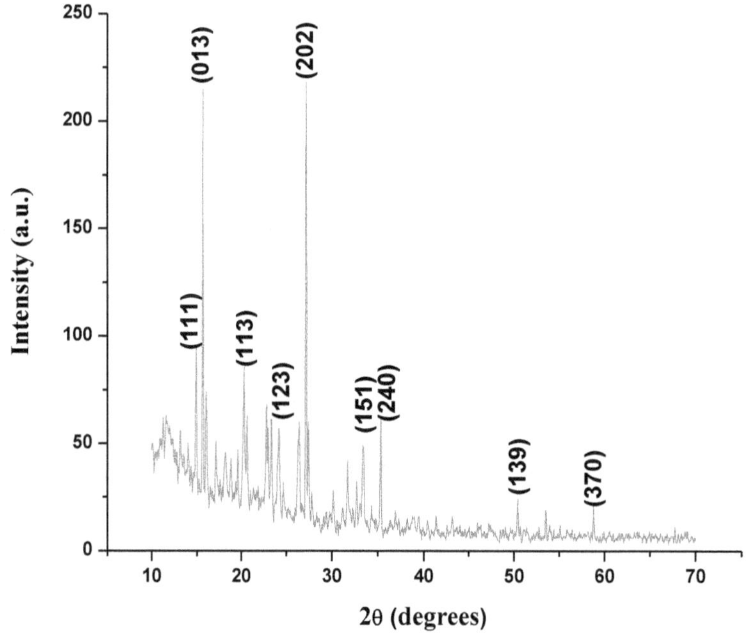


**Figure 3.5 Powder X-ray diffractogram of DPGTM**

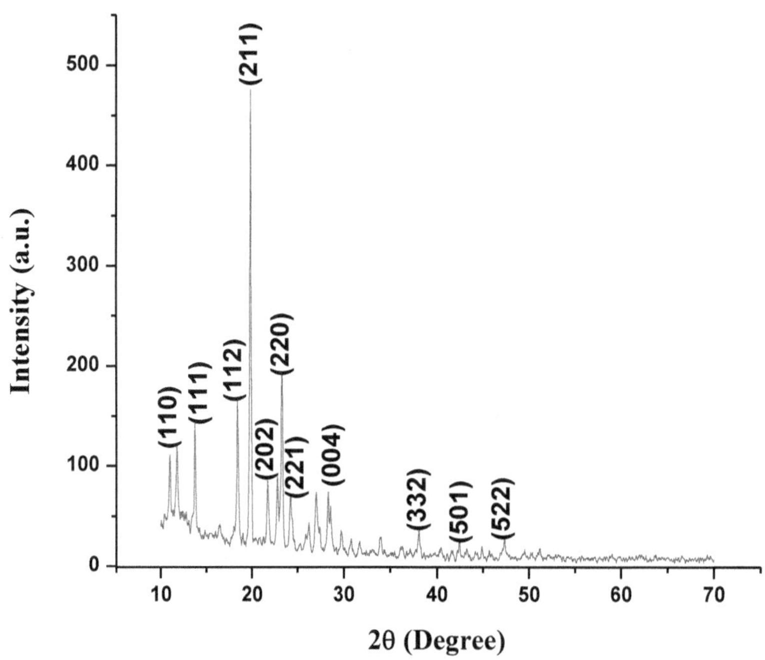


**Figure 3.6 Powder X-ray diffractogram of DPGP**

## 3.4 VIBRATIONAL SPECTROSCOPIC ANALYSIS

Vibrational spectroscopy is an important tool in the elucidation of molecular structure. The spectral bands can be assigned to different vibrational modes of the molecule. The various functional groups present in the molecule can be assigned by the comparison of spectra with characteristic functional group frequencies. As the positions of the bands are directly related to the strength of the chemical bond, a large number of investigation including intermolecular interactions and chemical kinetics can be carried out using this branch of spectroscopy. Infrared and Raman spectroscopies are two complimentary techniques to identify the functional groups which are adopted in the present study.

Intense Raman scattering occurs from vibrations which cause a change in the polarizability of the electron cloud around the molecule. Usually, symmetric vibrations cause the largest changes and give the greatest scattering. This contrasts with infrared absorption where the most intense absorption is caused by a change in dipole and hence asymmetric vibrations which cause this are the most intense. As will be seen later, not all vibrations of a molecule need, or in some cases can, be both infrared and Raman active and the two techniques usually give quite different intensity patterns. As a result the two are often complementary and used together, give a better view of the vibrational structure of a molecule. One specific class of molecule provides an additional selection rule. In a centrosymmetric molecule, no band can be active in both Raman scattering and infrared. But however in case of a non-centrosymmetric molecule almost all bands are active in both Raman scattering and infrared. This is called the mutual exclusion rule. Even though almost all bands are active in both Raman and infrared in case of a non-centrosymmetric molecule there are a few bands which does not appear in both and this may be attributed to the fact, the electron cloud will alter as the positive nuclei change position and depending on the nature of the change, this can cause a change of dipole moment or polarization. For example, the symmetrical stretch causes large polarization changes and hence strong Raman scattering with weak or no dipole change and hence weak or no infrared absorption. The deformation mode causes a dipole change but little polarization change and hence strong infrared absorption and weak or non-existent Raman scattering.

### 3.4.1 Fourier Transform Infrared (FT-IR) Spectral Studies

Infrared spectroscopy is effectively used to determine the molecular structure and the identification of functional groups in the synthesized compounds (Nakamoto, 1986). Fourier Transform Infrared (FT-IR) spectroscopy is one of the most powerful analytical technique which propose the possibility of chemical identification. An infrared spectrum represents a fingerprint of a sample with absorption peaks which correspond to the frequencies of vibrations between the bonds of the atoms making up the material. Because each material has unique combination of atoms, no two compounds produce the similar infrared spectrum.

### 3.4.2 Fourier Transform Raman Spectral Studies

Fourier Transform Raman (FT-Raman) spectroscopy has the same advantages as other applications of interferometric technique. Raman spectroscopy is the measurement, as a function of wavenumber, of the inelastic light scattering that results from the excitation of vibrations in molecular and crystalline materials. The excitation source is a single line of a continuous gas laser, which permits optical microscope optics to be used for measurement of samples down to a few micro meters. Raman spectroscopy is sensitive to molecular and crystal structure.

Raman spectroscopy is primarily a structural characterization tool. The spectrum is more sensitive to the lengths, strengths and arrangement of bonds in a material than it is to the chemical composition. The Raman spectrum of crystals likewise responds more to details of defects and disorder than to trace impurities and related chemical imperfections. From the knowledge of the crystal structure it is possible to calculate selection rules for each orientation position and thus gain considerable insight in to the vibrational motions of the crystal. The interpretation of such spectra, which show a lot of details, goes well beyond characterization applications.

### 3.4.3 FT-IR and FT-Raman Spectral Analyses of N,N'-Diphenylguanidine Based Single Crystals

In order to analyze the presence of functional groups in the grown crystals, they were subjected to FT-IR analysis using PERKIN-ELMER spectrometer using the KBr pellet technique with a resolution of 1.0 $cm^{-1}$. The samples were prepared by pressed pellet technique. In this technique, a small amount of the ground solid sample is mixed with about 100 times the weight of potassium bromide (KBr). The finely grounded mixture is pressed under a very high pressure of about 3 MPa in a press to form a small pellet of diameter of about 1 cm and 1.2 mm thickness. These pellets were exposed to a source of continuous infrared radiation. The spectrophotometer dispenses the radiation into a spectrum and the detector converts the radiant energy into an electrical signal, which is amplified electronically and recorded by a chart recorder.

The identities, surrounding environments and concentrations of chemical bonds that are present can be determined. A relatively recent development is the adaptation of Fourier transform spectrometers commonly used for infrared absorption spectroscopy to Raman spectroscopy. Usually, particle size has relatively little effect on Raman line shapes unless the particles are extremely small, less than 100 nm. For this reason, high-quality Raman spectra can be obtained from powders by simply reflecting the laser beam from the specimen. FT-Raman spectrum was recorded using FT-RAMAN spectrometer (BRUKER RFS 27: Stand-alone FT-Raman Spectrometer) with a resolution of 2.0 $cm^{-1}$. The FT-IR and FT-Raman spectra of the N,N'-diphenylguanidine series of compounds were recorded in the range 4000 - 400 $cm^{-1}$.

*N,N'-Diphenylguanidinium Nitrate (DPGN)*

The spectral analyses of DPGN was carried out on the characteristic group vibrations of guanidinium groups, nitrate ions and phenyl rings. The FT-IR and FT-Raman spectra of DPGN compound are presented in figures 3.7 and 3.8. The vibrational frequency of various functional groups of DPGN and the frequency assignments are presented in Table 3.5.

**Table 3.5 Vibrational band assignments of DPGN**

| Wavenumber ($cm^{-1}$) | | |
|---|---|---|
| FT-IR | FT-Raman | Vibrational assignments |
| 3352 | - | $NH_2$ Asymmetric stretching of guanidinium |
| 3223 | - | $NH_2$ Symmetric stretching guanidinium |
| 3070 | 3071 | C-H stretching of the phenyl rings |
| 1645 | 1662 | Asymmetric C-N stretching of guanidinium |
| 1582 | 1547 | $NH_2$ bending vibrations of guanidinium |
| 1385 | - | N-O stretching of nitrate |
| 1307 | 1306 | C-C stretching of phenyl rings |
| 1158 | 1159 | In-plane C-H bending of phenyl ring |
| 1092 | 1087 | $NH_2$ rocking in-plane vibrations of guanidinium |
| 1002 | 995 | C-N symmetric stretching of guanidinium |
| 883 | 884 | Out-of plane C-H bending of phenyl ring |
| 825 | - | N-O out of plane bending of nitrate |
| 754 | 754 | N-O in plane bending of nitrate |

***Hydrogen Bond***

The very broad nature of the spectrum in the region above 3000 $cm^{-1}$ in IR clearly shows the strong hydrogen bond interactions between the anions and cations. The hydrogen bond interaction paves way for the second harmonic generation (Saravanakumar and Murugakoothan, 2014).

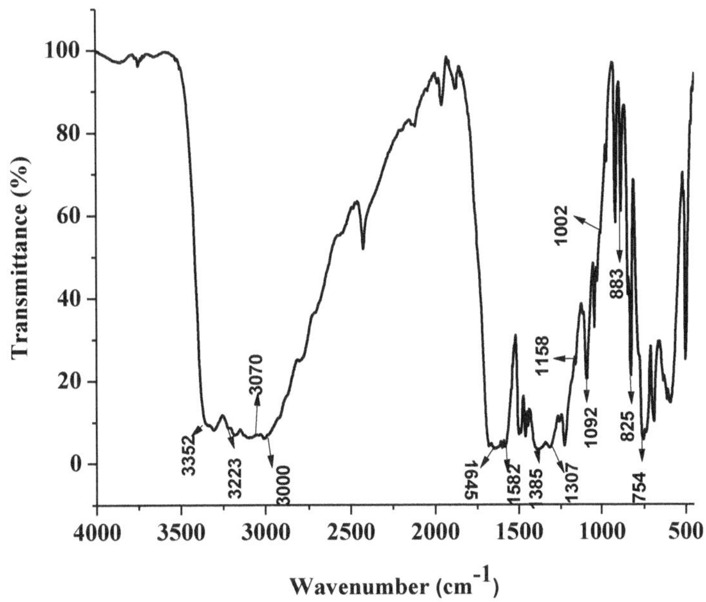


**Figure 3.7 FT-IR spectrum of DPGN**

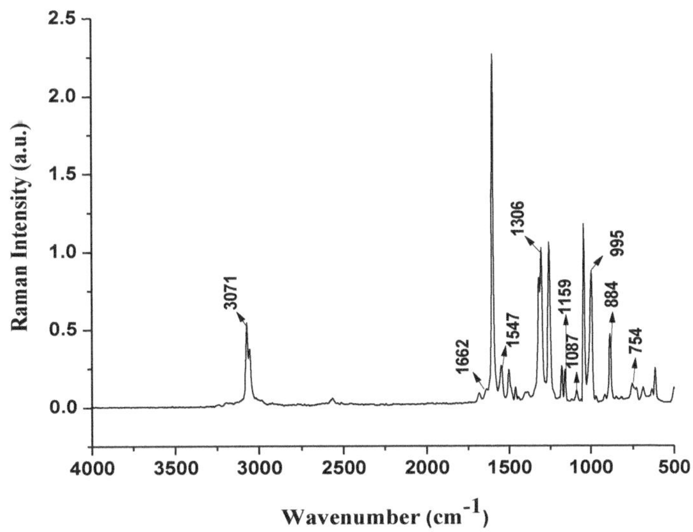


**Figure 3.8 FT-Raman spectrum of DPGN**

### *Guanidinium Cation Vibrations*

The high frequency absorption peak at 3352 $cm^{-1}$ and 3223 $cm^{-1}$ in IR are assigned to $NH_2$ asymmetric and symmetric stretching mode respectively. The characteristic band observed at 995 $cm^{-1}$ in the Raman spectrum is assigned to the symmetric C-N stretching vibrations but its IR counterpart is observed as very weak intensity at 1002 $cm^{-1}$. The band observed in IR spectrum at 1645 $cm^{-1}$ is assigned to the asymmetric C-N stretching mode (Drozd et. al., 2005c). In Raman spectrum a weak band corresponding to this type of vibration is observed at 1662 $cm^{-1}$. The very strong band at 1582 $cm^{-1}$ in the IR spectrum arises from the $NH_2$ bending vibrations. The Raman counterpart for these vibrations appear at 1547 $cm^{-1}$. The bands at 1092 $cm^{-1}$ and 1087 $cm^{-1}$ in the IR and Raman spectra correspond to the $NH_2$ rocking in-plane vibrations.

### *Phenyl Rings Vibrations*

The phenyl ring is a cyclic group of atoms with formula $C_6H_5$. Phenyl groups have six carbon atoms bonded together in a hexagonal planar ring, five of which are bonded to individual hydrogen atoms, with the remaining carbon bonded to a substituent. The phenyl group vibrations are assigned on the basis of C-C, C-H vibrations. Though the title compound has two mono substituted phenyl rings there are no separate vibrations for the both. The C-H stretching frequencies of the mono substituted benzene are expected in the region 3010 – 3120 $cm^{-1}$ (James et. al., 2007). The very strong band at 3070 $cm^{-1}$ in the IR spectrum is attributed to the C-H stretching of the phenyl rings. The Raman counterpart of this vibration occurs at 3071 $cm^{-1}$. The strong intensity band at 1158 $cm^{-1}$ in IR spectrum is assigned to the in-plane C-H bending of monosubstituted benzene ring and its Raman counterpart is found to occur at 1159 $cm^{-1}$ with strong intensity. The allowed vibrational frequencies arising from the C-H out of plane bending vibrations of the monosubstituted benzene ring are expected in the region 1000 – 650 $cm^{-1}$ (James et. al., 2007). The medium band observed at 883 $cm^{-1}$ in IR is assigned to the C-H out of plane bending vibrations. The Raman counterpart arising from this vibration is found to occur with a strong intensity at 884 $cm^{-1}$. The C-C stretching modes of vibrations are expected in the range 1300 $cm^{-1}$ (James et. al., 2007). The strong intensity band at 1306 $cm^{-1}$ in Raman is assigned to the C-C stretching mode of vibration. The IR counterpart for this mode is found to occur at 1307 $cm^{-1}$ with very strong intensity.

***Nitrate Anion Vibrations***

The presence of nitrate anions in the title compound can be confirmed by the presence of characteristic vibrations arising due to N-O stretching and the in-plane and out of plane bending vibrations of nitrate ions. The N-O stretch of nitrate groups appears as an intense band in the region between 1400 and 1340 $cm^{-1}$. The in-plane and out of plane bending vibrations of the nitrate groups appear in the region around 720 $cm^{-1}$ and 840 – 810 $cm^{-1}$ respectively (Smith, 1999). The very strong intensity band at 1385 $cm^{-1}$ in IR spectrum is assigned to the N-O stretching. The very strong IR intensity at 754 $cm^{-1}$ and weak band at 754 $cm^{-1}$ in Raman is assigned to the N-O in plane bending vibrations. The strong band occurring at 825 $cm^{-1}$ is arising due to N-O out of plane bending vibrations. Even though the IR and Raman spectra of DPGN compound consists of the characteristic vibrations of both the cation and anion it is N,N'-Diphenylguanidinium Nitrate.

***N,N'-Diphenylguanidinium Hydrogen (+) -L- Tartrate Monohydrate (DPGTM)***

The chemical bonding and the presence of various expected functional groups were identified by recording the FT-IR and FT-Raman spectra. The bands observed in the measured region arise from the vibrations of guanidinium cations, hydrogen bonds, tartrate anions, phenyl rings and water of crystallization. The FT-IR and FT-Raman spectra of DPGTM are shown in figures 3.9 and 3.10. The vibrational frequency of various functional groups of DPGTM and the frequency assignments are presented in Table 3.6.

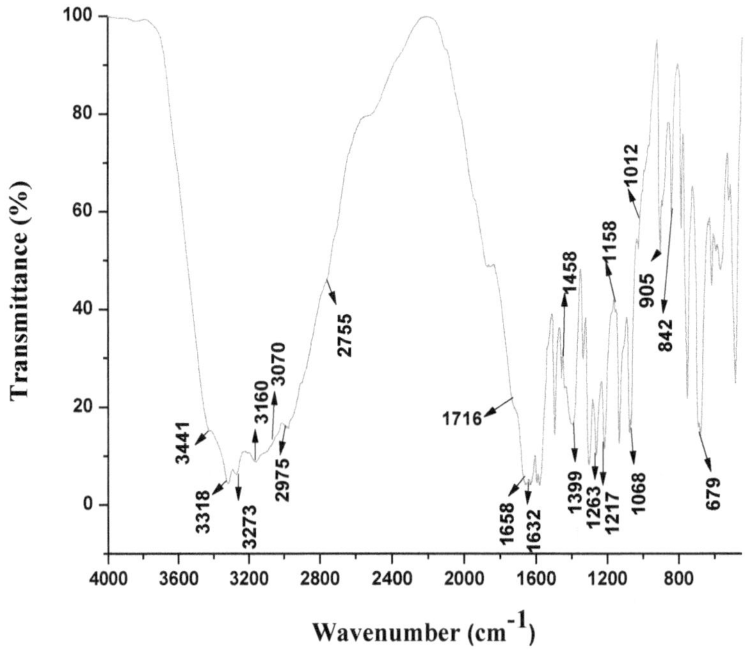


**Figure 3.9 FT-IR spectrum of DPGTM**

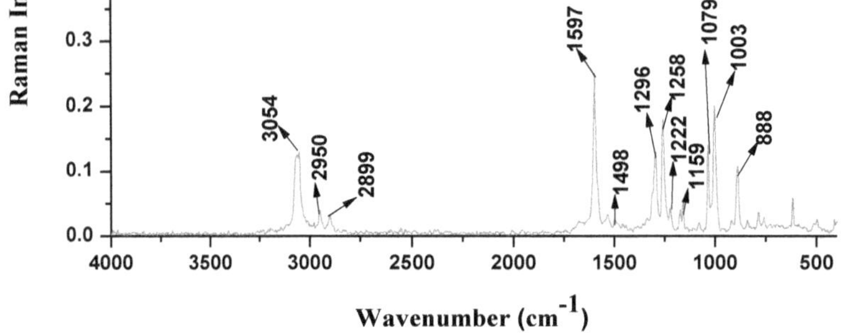


**Figure 3.10 FT-Raman spectrum of DPGTM**

### *Water of Crystallization*

The very strong bands present at 3273 $cm^{-1}$ and 3160 $cm^{-1}$ are assigned to the asymmetric and symmetric stretching of OH molecules of water of crystallization. The Raman counterpart of these vibrations are found to occur at 2950 $cm^{-1}$ and 2899 $cm^{-1}$ respectively. The very strong band at 1658 $cm^{-1}$ is assigned to bending vibrations of water (Saravanakumar and Murugakoothan, 2015a).

### *Hydrogen Bonding Vibrations*

The band observed at 2975 $cm^{-1}$ in IR spectrum and the broad nature of the spectra above this region clearly indicates the presence of strong hydrogen bonding which is responsible for second harmonic generation.

### *Guanidinium Cation Vibrations*

The very strong bands observed at 3441 $cm^{-1}$ and 3318 $cm^{-1}$ in the IR spectrum are assigned to the $NH_2$ asymmetric and symmetric stretching vibrations respectively (Saravanakumar and Murugakoothan, 2015a). The weak band in IR spectrum observed at 1012 $cm^{-1}$ is assigned to the symmetric stretching of C-N. The Raman counterpart for this vibration occur at 1034 $cm^{-1}$. The very strong band at 1632 $cm^{-1}$ in IR spectrum is assigned to the asymmetric C-N stretching of guanidinium fragments. The Raman counterpart of C-N asymmetric stretching vibrations is found to occur at 1597 $cm^{-1}$.

### *Phenyl Rings Vibrations*

The C-H stretching frequencies of the monosubstituted benzene are expected in the region 3010 – 3120 $cm^{-1}$ (Saravanakumar and Murugakoothan, 2014). The very strong band at 3070 $cm^{-1}$ in the IR spectrum is attributed to the C-H stretching of the phenyl rings. However for Raman counterpart this vibration occurs at 3054 $cm^{-1}$. The very strong intensity band at 1077 $cm^{-1}$ and 1263 $cm^{-1}$, the strong intensity band at 1158 $cm^{-1}$ is assigned to the in-plane C-H bending of monosubstituted benzene ring. The In-plane C-H bending vibrations occurs at

1079 $cm^{-1}$, 1258 $cm^{-1}$ and 1158 $cm^{-1}$ in Raman spectrum. The allowed vibrational frequencies arising from the C-H out of plane bending vibrations of the monosubstituted benzene ring are expected in the region 1000 – 650 $cm^{-1}$. The weak bands observed at 842 $cm^{-1}$, 893 $cm^{-1}$ and 905 $cm^{-1}$ and very strong bands at 679 $cm^{-1}$ and 692 $cm^{-1}$ are assigned to the C-H out of plane bending vibrations. The Raman counterpart for this mode of vibrations occurs at 840 $cm^{-1}$, 888 $cm^{-1}$ and 1003 $cm^{-1}$ respectively. The C-C stretching mode of vibrations is expected in the range 1300 $cm^{-1}$ as low or medium to high (Saravanakumar and Murugakoothan, 2014). The very strong band at 1303 $cm^{-1}$ is assigned to this vibration. The Raman counterpart of C-C stretching mode of vibrations occur at 1296 $cm^{-1}$.

### *Tartrate Anion Vibrations*

The presence of C=O group of tartrate is confirmed by the peak at 1716 $cm^{-1}$ in IR spectrum. The asymmetric stretching of $COO^-$ group in IR spectrum is assigned at 1632 $cm^{-1}$. In IR spectrum the C-H vibration of tartrate occurs at 2755 $cm^{-1}$. The deformation vibrations of the OH and CH of tartrate are mixed and are observed at 1458 $cm^{-1}$ and 1399 $cm^{-1}$ in IR spectrum. The Raman counterpart of the mixed vibration occurs at 1498 $cm^{-1}$. The $COO^-$ stretching is observed as very strong band at 1217$cm^{-1}$ in IR spectrum. The Raman counterpart of this band occurs at 1222 $cm^{-1}$ (Saravanakumar and Murugakoothan, 2015a). The C–O vibration of tartrate is observed at 1068 $cm^{-1}$ in IR spectrum. The C–O vibration of tartrate occur at 1079 $cm^{-1}$ in Raman spectrum. The C-C stretching modes of tartrate is found to be mixed with the stretching modes of phenyl rings and also occurs at the same modes of vibrations as that of the phenyl rings. As these spectra carries the vibrations of both N,N'- Diphenylguanidine and tartaric acid, the crystal must have both compounds together.

**Table 3.6 Vibrational band assignments of DPGTM**

| Wavenumber ($cm^{-1}$) | | |
|---|---|---|
| FT-IR | FT-Raman | Vibrational Assignments |
| 3441 | - | $NH_2$ Asymmetric stretching of guanidinium |
| 3318 | - | $NH_2$ Symmetric stretching of guanidinium |
| 3273 | 2950 | OH Asymmetric stretching of water of crystallization |
| 3160 | 2899 | OH symmetric stretching of water of crystallization |
| 3070 | 3054 | C – H stretching of phenyl rings |
| 2755 | - | C – H vibrations of tartrate |
| 1716 | - | C = O group of tartrate |
| 1632 | - | Asymmetric stretching of $COO^-$ of tartrate |
| 1597 | 1597 | C-N Asymmetric stretching of guanidinium |
| 1458 | 1498 | Deformation of OH & CH groups of tartrate |
| 1303 | 1296 | C-C stretching of phenyl rings |
| 1263 | 1258 | In-plane C-H bending of Phenyl ring |
| 1217 | 1222 | $COO^-$ stretching of tartrate |
| 1158 | 1159 | In-plane C-H bending of Phenyl ring |
| 1077 | 1079 | In-plane C-H bending of Phenyl ring |
| 1068 | 1079 | C–O vibration of tartrate |
| 1012 | 1034 | Symmetric stretching of C-N of guanidinium |
| 905 | 1003 | Out-of-plane C-H bending of Phenyl ring |
| 893 | 888 | Out-of-plane C-H bending of Phenyl ring |
| 842 | 840 | Out-of-plane C-H bending of Phenyl ring |

### *N,N′-Diphenylguanidinium Dihydrogen Phosphite (DPGP)*

The spectral analyses of DPGP was carried out on the characteristic vibrations of guanidinium group, phosphite ion and phenyl rings. The FT-IR and FT-Raman spectra of DPGP compound are presented in figures 3.11 and 3.12 respectively. The vibrational frequency of various functional groups of DPGP and the frequency assignments are presented in Table 3.7.

### *Guanidinium Cation Vibrations*

The high frequency absorption peak at 3440 $cm^{-1}$ and 3285 $cm^{-1}$ in IR spectrum are assigned to $NH_2$ asymmetric and symmetric stretching modes of vibration respectively. The characteristic band observed at 998 $cm^{-1}$ in the Raman spectrum is assigned to the symmetric C-N stretching vibrations. The IR counterpart is observed as very strong band at 999 $cm^{-1}$. The band observed in IR spectrum at 1660 $cm^{-1}$ is assigned to the asymmetric C-N stretching mode. The very strong band at 1573 $cm^{-1}$ in the IR spectrum arises from the $NH_2$ bending vibrations. The Raman counterpart for these vibrations appears at 1597 $cm^{-1}$. The presence of characteristic bands at 1062 $cm^{-1}$ in the IR spectrum and 1032 $cm^{-1}$ in Raman spectrum corresponds to the $NH_2$ rocking in-plane mode of vibrations (Saravanakumar et. al., 2015b).

### *Phenyl Rings Vibrations*

The C-H stretching frequencies of the monosubstituted benzene are expected in the region 3010 – 3120 $cm^{-1}$. The very strong band at 3070 $cm^{-1}$ in the IR spectrum is attributed to the C-H stretching of the phenyl rings. The Raman counterpart of this vibration occurs at 3071 $cm^{-1}$. The very strong intensity band at 1138 $cm^{-1}$ in IR spectrum is assigned to the in-plane C-H bending of monosubstituted benzene ring. The Raman counterpart is found to occur at 1153 $cm^{-1}$. The allowed vibrational frequencies arising from the C-H out of plane bending vibrations of the monosubstituted benzene ring are expected in the region 1000 – 650 $cm^{-1}$. The strong band observed at 885 $cm^{-1}$ in IR spectrum is assigned to the C-H out of plane bending vibration. The Raman counterpart arising from this vibration is found to occur at 887 $cm^{-1}$. The C-C stretching modes of vibrations are expected in the range 1300 $cm^{-1}$. The strong intensity band at 1288 $cm^{-1}$ in Raman is assigned to the C-C stretching mode of vibration. The IR counterpart for this mode is found to occur at 1312 $cm^{-1}$ with very strong intensity.

***Phosphite Anion Vibrations***

The presence of phosphite anion in the title compound can be confirmed by the presence of characteristic vibrations arising due to P-O, P-H, P-O-H ions. The P-H stretch of phosphite group appears as a sharp band in the region between 2440 and 2275 $cm^{-1}$ (Colthup et. al., 1964). The sharp band observed at 2372 $cm^{-1}$ in the IR spectrum corresponds to the P-H stretching of phosphite ion. The counterpart for this vibration in the Raman spectrum with a sharp band is found at 2373 $cm^{-1}$. The P-H wagging vibration found to occur in the range 990 – 965 $cm^{-1}$ (Colthup et. al., 1964). The P-H wagging vibrations of phosphite ion in the title compound occurs at 999 and 998 $cm^{-1}$ respectively in the IR and Raman spectrum. The stretching of P = O bond gives rise to a strong band generally at 1320 – 1140 $cm^{-1}$ (Colthup et. al., 1964). A very strong band at 1236 $cm^{-1}$ in the IR spectrum corresponds to P = O stretching of phosphite ion. However the Raman counterpart for this vibration occur at 1262 $cm^{-1}$. Mostly phosphorous acid which have a P-OH group have a strong band at 1040 – 910 $cm^{-1}$, probably involving the stretching of the P-OH bonds. The OH stretching of the P-OH group gives rise to medium intensity broad bands at 2700 – 2550 $cm^{-1}$ and 2350 – 2100 $cm^{-1}$ (Colthup et. al., 1964). These have been assigned to the OH stretching vibration of the phosphinic acid. The characteristic bands found at 925, 952, and 1032 $cm^{-1}$ in the Raman spectrum can be attributed to the OH stretching of P-OH group. The band in IR spectrum at 2400 $cm^{-1}$ is attributed to the OH stretching of phosphinic acid and also to the hydrogen bonding between N,N'-diphenylguanidine and phosphinic acid which paves the way for second harmonic generation of title compound. It can be concluded that DPGP consists of the characteristic vibrations of both the cation and anion and thus it can be claimed that the title compound are not separate entities but it is N,N'-diphenylguanidinium dihydrogen phosphite.

**Table 3.7 Vibrational band assignments of DPGP**

| Wavenumber ($cm^{-1}$) | | |
|---|---|---|
| FT-IR | FT-Raman | Vibrational Assignments |
| 3440 | - | $NH_2$ Asymmetric stretching of guanidinium |
| 3285 | - | $NH_2$ Symmetric stretching of guanidinium |
| 3070 | 3071 | C-H stretching of the phenyl rings |
| 2400 | - | O – H stretching of phosphite |
| 2372 | 2373 | P – H stretching of phosphite |
| 1660 | - | Asymmetric C-N stretching of guanidinium |
| 1573 | 1597 | $NH_2$ bending vibrations of guanidinium |
| 1312 | 1288 | C-C stretching of phenyl rings |
| 1236 | 1262 | P = O stretching of phosphite |
| 1138 | 1153 | In-plane C-H bending of Phenyl ring |
| 1062 | 1032 | $NH_2$ rocking in-plane vibrations of guanidinium |
| 999 | 998 | C-N symmetric stretching of guanidinium |
| - | 952 | O-H stretching of phosphite |
| - | 925 | O-H stretching of phosphite |
| 885 | 887 | Out-of plane C-H bending of phenyl ring |

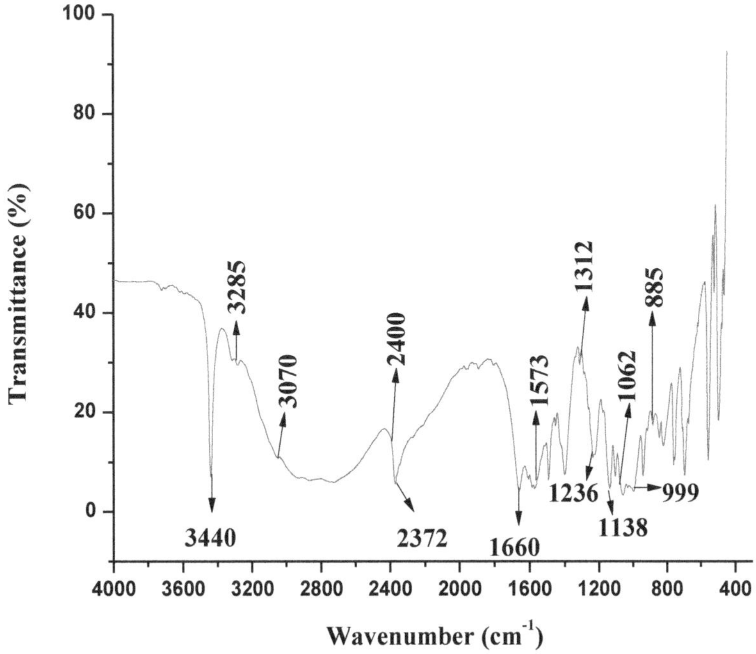


**Figure 3.11 FT-IR spectrum of DPGP**

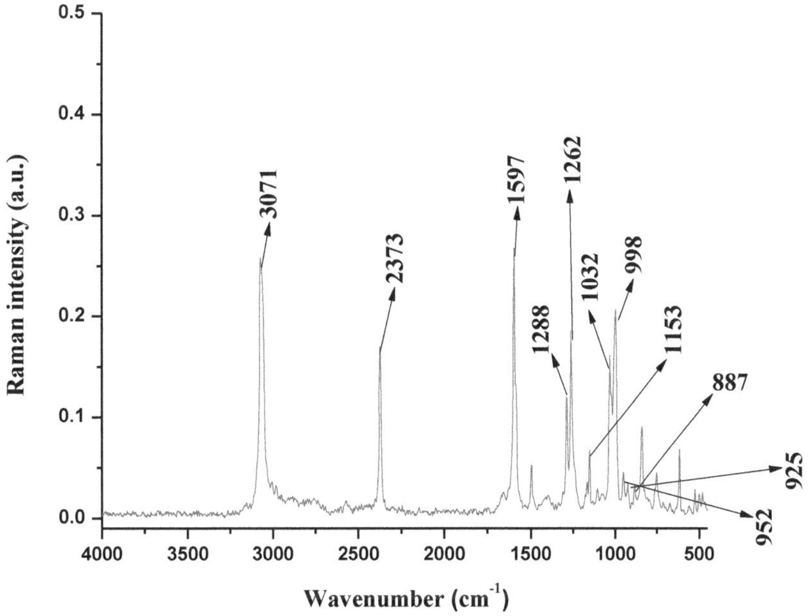


**Figure 3.12 FT-Raman spectrum of DPGP**

From these spectral analyses, it can be confirmed that the expected various functional groups are present and the formation of N,N'-diphenylguanidine based NLO compounds are due to the strong hydrogen bonding interactions between the donor and acceptor which is the necessary condition and pathway for the second harmonic generation (SHG) of the title compounds.

## 3.5 CONCLUSION

The structural and spectral analyses are unique techniques which are used to reveal the structure and modes of vibrations of the molecules. Overall results presented here suggest that there is a stabilizing interaction conferred by the pair of ions of the compounds studied in the present work. The single crystal XRD study was carried out to find unit cell parameters and the crystal system of grown N,N'-diphenylguanidine series compounds. The X- ray diffraction studies reveal that the grown DPGN and DPGTM crystals belong to orthorhombic system while DPGP belong to tetragonal crystal system. The optimized molecular geometry of the title compounds was carried out using GAUSSIAN 03W computational software employing Density Functional Theory (DFT). The optimized molecular geometry represents an isolated single molecule under ideal conditions with a stationary point at the potential energy surface; the convergence was confirmed by observing no imaginary vibrational wavenumbers obtained in the calculation and thereby claiming that the title compounds are stable. The unit cell dimensions evaluated from powder XRD are in good agreement with that of single crystal XRD analysis and also show the good crystallinity of the crystals. The modes of vibration of the molecules and the presence of all functional groups have been identified using FT-IR and FT-Raman techniques.

# CHAPTER – IV

# LINEAR AND NONLINEAR OPTICAL STUDIES OF N,N'-DIPHENYLGUANIDINE BASED NLO SINGLE CRYSTALS

## 4.1 INTRODUCTION

Nonlinear optical (NLO) materials find extensive optoelectronic applications, such as optical frequency conversion, optical data storage and optical switches in the inertial confined laser fusion systems. The common wisdom has been that an optical material should have large charge transfer and optical transparency with less dislocation density. Even though varieties of NLO materials are available, their applications are limited due to the physico-chemical properties (Antony et. al., 2009; Oleg Kolosov et. al., 1995). Organic materials exhibit higher nonlinear optical efficiency when compare to inorganic. The high second harmonic generation (SHG) nonlinearity and transparency in ultra violet and visible regions are required for various devices in the field of optoelectronics and photonics.

Complexes of organic material with inorganic acids are promising materials for optical second harmonic generation (SHG) as they tend to combine the features of organic with that of inorganic materials. Among the different types of NLO materials, organic materials are shown to be superior to their inorganic counter parts in terms of synthesis, molecular engineering and much faster optical nonlinearities. It is well known fact that purely organic compounds lack mechanical strength for practical uses and it is difficult to grow single crystals of sufficient quality. For this reason the crystal engineering strategy of combining organic and inorganic moieties was proposed with the objective of building more cohesive crystalline structures. These systems exhibit high nonlinear optical efficiency in the visible region, good thermal and mechanical stability compared with those of purely organic crystals (Saravana Kumar et. al., 2015b).

Second-order nonlinear optical materials capable of efficient frequency conversion of infrared or visible laser radiation to visible or ultraviolet (UV) wavelengths are of considerable interest in the field of telecommunications, high density optical recording, color display, medical diagnostics etc. Therefore, presently there is a need to produce high efficiency NLO materials capable of generating blue light by second harmonic generation (Chemla and Zyss, 1987). Materials with second-order optical nonlinearities and short transparency cutoff wavelength are needed to realize many of these applications. Recently, it has been shown that some hydrogen-bonded crystals are well suited for such purposes. Several researchers have investigated the influence of the hydrogen bonds on nonlinear optical materials.

The guanidinium ion can form a broad family of hydrogen-bonded crystals. The presence of potential donor sites for hydrogen-bonding interactions and delocalized electron systems has made guanidinium compounds potentially interesting material for non-linear optical applications (Zyss et. al., 1993). The family of guanidinium complexes is intensively investigated as potential materials for nonlinear optical efficiency with ferroic phase transitions (Marek Szafranski and Andrzej Katrusiak, 2006). In recent years, efforts have been made on guanidine and its derivatives of mixed organic and inorganic acid crystals to improve the chemical stability, laser damage threshold and nonlinear optical properties.

In this chapter, the optical responses of N,N'-diphenylguanidine based crystals were analyzed through UV-vis spectral technique and the nonlinear optical response was analyzed using Kurtz and Perry (Kurtz and Perry, 1968) powder technique and laser damage threshold studies. Theoretical investigations was carried out for the title crystals in order to analyze the HOMO-LUMO energy gap from the optimized molecular geometry to reveal the electronic transitions. Nonlinear optical parameters, such as dipole moment, polarizability and first order hyperpolarizability were obtained at molecular level to understand the NLO behaviour.

## 4.2 LINEAR OPTICAL ANALYSIS

### 4.2.1 Introduction

Ultraviolet-visible spectroscopy involves the spectroscopy of photons in the UV-visible region. This means it uses light in the visible and adjacent (near ultraviolet and near infrared) regions. The absorption in the visible region directly affects the color of the chemicals involved. In this region of the electromagnetic spectrum, molecules undergo electronic transitions. In this technique absorption is attributed to a process in which the outer electrons of atoms or molecules absorb radiant energy and undergo transitions to higher energy levels. UV-vis spectroscopy is used in the quantitative determination of solutions of transition metal ions and highly conjugated organic compounds (Dyke et. al., 1978).

### 4.2.2 Quantum Theory

Quantum theory is necessary to gain an understanding of the origin of practical absorption spectrometry. For this purpose, it is better to think of radiation as a stream of particles known as photons instead of waves. Atoms and molecules exist in a number of defined energy states or levels and a change of level requires the absorption or emission of an integral number of a unit of energy called as a quantum or a photon. The energy of a photon absorbed or emitted during a transition from one molecular energy level to another is given by the equation:

$$E = hv \qquad (4.1)$$

where $h$ the Planck's constant and $v$ the frequency of the photon.

But, $$c = \nu\lambda \qquad (4.2)$$

Therefore, $$E = \frac{hc}{\lambda} \qquad (4.3)$$

A molecule of any substance has an internal energy, which can be considered as the sum of the energy of its electrons, the energy of vibration between its constituent atoms and the energy associated with rotation of the molecule. The electronic energy levels of simple molecules are widely separated and usually only the absorption of a high energy photon, i.e., one of the very short wavelengths can excite a molecule from one level to another.

In complex molecules, the energy levels are more closely spaced and photons of near ultraviolet and visible light can affect the transition. These substances, therefore, will absorb light in some areas of the near ultraviolet and visible regions. The vibrational energy states of the various parts of a molecule are much closer than the electronic energy levels and thus photons of lower energy (longer wavelength) are sufficient to bring about vibrational changes. Light absorption due to vibrational changes occurs in the infrared region. The rotational energy states of molecules are so closely spaced than vibrational states, hence light in the far infrared and microwave regions of the electromagnetic spectrum has enough energy to cause these small changes.

### 4.2.3 Instrumentation

The spectrometers for measuring absorption of radiation are made up of one or more,

(i). radiation source,
(ii). wavelength selectors,
(iii). sample containers,
(iv). radiation detectors and
(v). signal processors and read out devices.

The hydrogen or deuterium discharge lamp operating at a low pressure produces continuous UV spectrum. Quartz or fused silica is used for the construction of cells or cuvettes that hold the sample and solvent as these are transparent in UV region. Diffraction grating or prism can be used as monochromators. Filter photometers allow a wide band of radiations to strike the detector thus avoid energy loss due to monochromators. Photovoltaic cells and photomultiplier tubes are used initially as detectors and now a-days photodiodes are used.

The modern spectrophotometers are computerized and the resulting detector output is digitalized and stored. Samples are scanned, the absorbance is calculated and the spectrum is displayed.

### 4.2.4 Single-Beam UV Spectrophotometer

Single-beam spectrophotometers are often sufficient for making quantitative absorption measurements in the UV-vis spectral region. The concentration of an analytic in solution can be determined by measuring the absorbance at a single wavelength and applying the Beer–Lambert law.

Single-beam spectrophotometers can utilize a fixed wavelength light source or a continuous source. The simplest instruments use a single wavelength light source, such as a light-emitting diode (LED), a sample container and a photodiode detector. Instruments with continuous source have a dispersing element and aperture or slit to select a single wavelength before the light passes through the sample cell. In this type of single-beam instrument, the instrument is calibrated with a reference cell containing only solvent to determine the light power ($P_0$) value necessary for absorbance measurement.

### 4.2.5 Double Beam UV Spectrophotometer

Most spectrophotometers are double-beam instruments. The primary source of light is divided in to two beams of equal intensity. One of the beams of selected monochromatic light passed through the sample material and the other beam of equal intensity is passed through the reference material. After the beam passed through the sample and reference materials, the intensities of the respective transmitted beams are compared over the whole wavelength range of the instrument. The detection electronics can then manipulate the measurements to give the absorbance or transmittance. The signal for the intensity of absorbance or transmittance versus corresponding wavelength is automatically recorded on the graph.

## 4.3 LINEAR OPTICAL STUDIES ON N,N'-DIPHENYLGUANIDINE BASED NLO SINGLE CRYSTALS

In order to reveal the optical properties of N,N'-diphenylguanidine series of single crystals, UV-vis transmission spectrum was recorded using VARIAN CARY 5E UV-vis spectrophotometer. The optical transmission range, percentage of transmission and cut-off wavelength are the most important optical parameters for laser frequency conversion applications.

The transmission range of the as grown N,N'-diphenylguanidine crystals, was determined by recording the optical transmission spectrum in the wavelength range 200 - 800 nm. The crystals of thickness 2 mm were used for this measurement. The as grown crystals of the title compounds was polished before carrying out the optical studies. The transmittance spectrum of DPGN, DPGTM and DPGP are shown in figure 4.1(a), 4.1(b) and 4.1(c) respectively. The cut-off wavelength of DPGN, DPGTM and DPGP were found to be 310 nm, 298 nm and 294 nm. It is evident that there is no remarkable absorption observed beyond the respective cut-off wavelengths till the entire region of the spectrum.

The percentage of transmission of about 56 %, 76 % and 70 % for DPGN, DPGTM and DPGP is attributed to better quality of the grown crystals. It is also interesting to note from the spectra that these crystals absorb UV radiation and hence the crystals can also be used as an effective UV shelter (Science News Online, 1998). As these crystals exhibit wide transmission range, it can be used for optical applications including the second harmonic generation of Nd: YAG laser of fundamental wavelength, $\lambda = 1064$ nm.

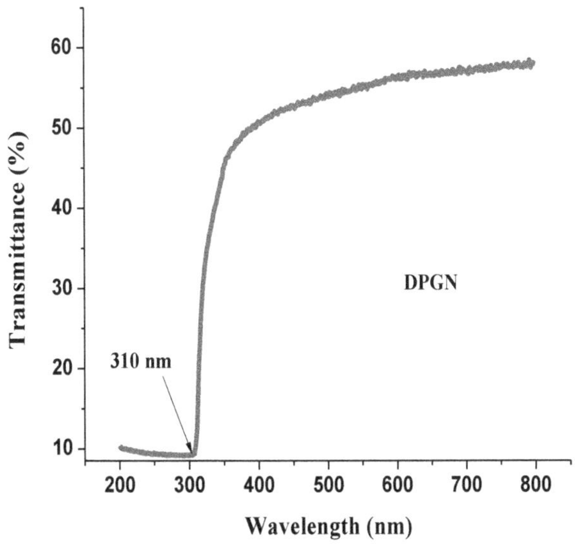


(a)

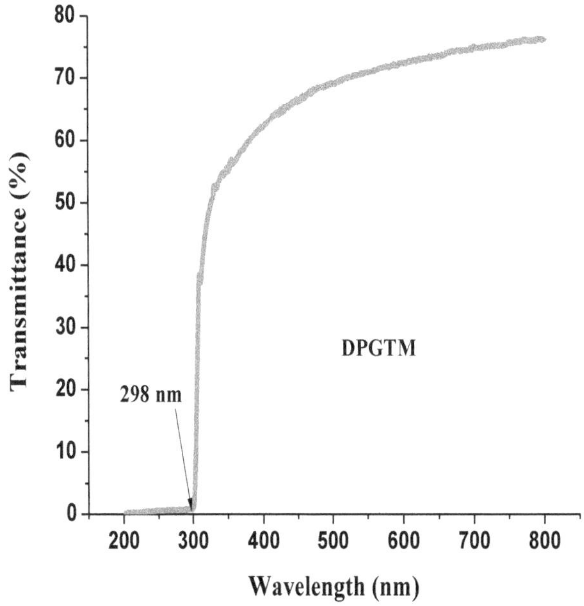


(b)

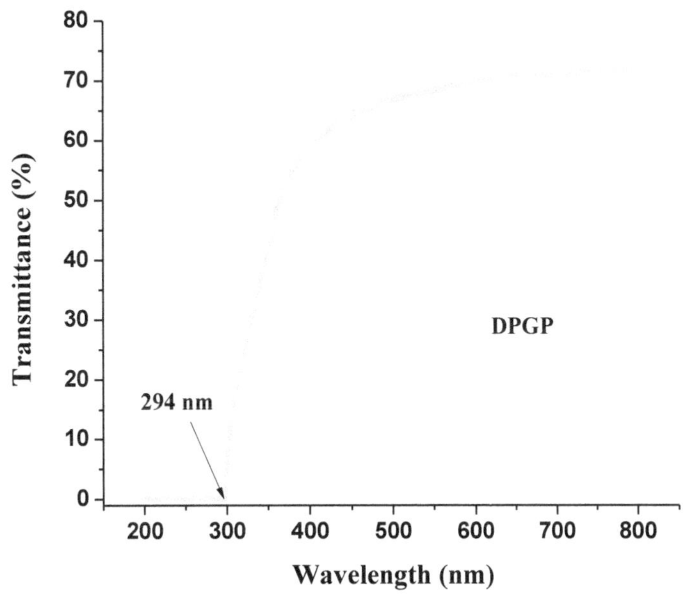


(c)

**Figure 4.1 Transmission spectra of (a) DPGN crystal, (b) DPGTM crystal and (c) DPGP crystal**

The transmission window in the visible region for the crystals of title compounds enables good optical transmission of the second harmonic frequencies of Nd:YAG lasers. The cut-off wavelength region of the grown N,N'-diphenylguanidine series of crystals is presented in Table 4.1.

**Table 4.1 The cut-off wavelength and band gap of the grown crystals**

| Crystals | Cut-off wavelength (nm) | Band gap energy (eV) |
|---|---|---|
| DPGN | 310 | 3.90 |
| DPGTM | 298 | 4.05 |
| DPGP | 294 | 4.14 |

### 4.3.1 Determination of Optical Band Gap

The dependence of optical absorption coefficient with the photon energy helps to study the band structure and the type of transition of electrons (Krishnan et. al., 2008). The optical absorption coefficient (α) is calculated from the transmittance data using the following relation:

$$\alpha = \frac{2.3026}{t} \log \left( {}^{1}/_{T} \right) \qquad (4.4)$$

where, $T$ the transmittance and $t$ the thickness of the crystal sample.

In the high photon energy region, the energy dependence of absorption coefficient suggests the occurrence of direct band gap of the crystal obeying the following equation for high photon energies $(h\nu)$.

The optical band gap ($E_g$) is evaluated from the transmission spectrum using the formula, (Ashour et. al., 1995).

$$\alpha h\nu = A\,(h\nu - E_g)^{1/2} \qquad (4.5)$$

where 'A' a constant, '$E_g$' the optical band gap, 'h' the Planck's constant and 'ν' the frequency of the incident photons.

The Tauc's graph plotted between the square of the product of absorption coefficient and the incident photon energy $(\alpha h\nu)^2$ with the photon energy $(h\nu)$ at room temperature shows a linear behavior that can be considered as evidence of the direct transition (Tauc et. al., 1966) and also it is clear that optical transition is an electronic transition from the valence band to conduction band with strong electronic absorption (Baccaro et. al., 1998).

The Tauc's plot between $(\alpha h\nu)^2$ and photon energy $(h\nu)$, to determine optical band gap value for the title crystals, were plotted and are shown in figure 4.2 (a), (b) and (c).

Ideally, the graph should be linear and the deviation from linearity at low incident photon energy can be attributed to the presence of irregularities and imperfections in the crystal. The optical band gap ($E_g$) is estimated by extrapolation of the linear portion near the onset of absorption edge to the energy axis, $(\alpha h\nu)^2 = 0$ (Davis and Mott, 1970). The band gap energies of the grown N,N'-diphenylguanidine series of crystals are presented in Table 4.1.

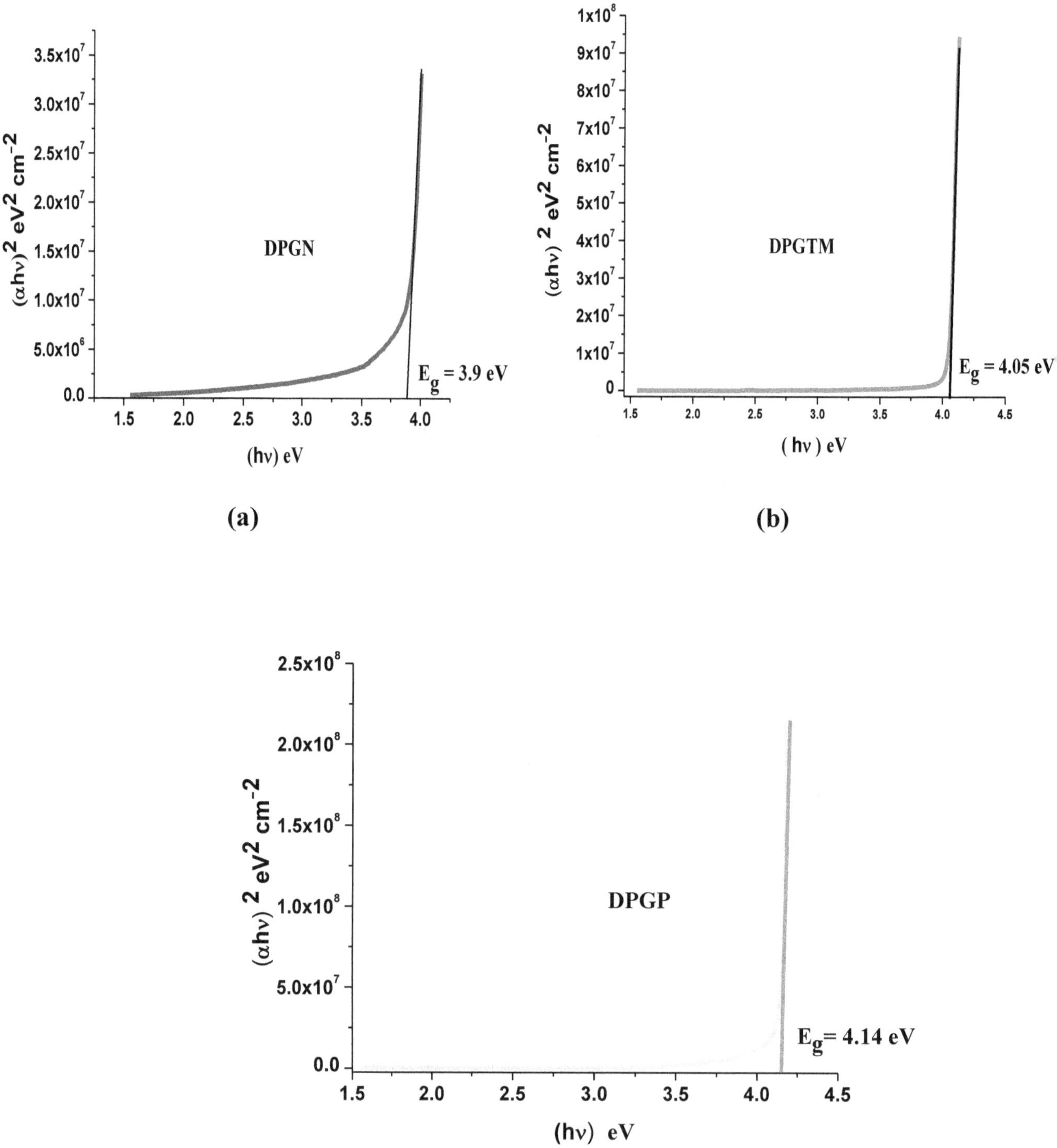


(a) (b) (c)

**Figure 4.2 Tauc's plot between $(\alpha h\nu)^2$ and photon energy $(h\nu)$ for (a) DPGN crystal, (b) DPGTM crystal and (c) DPGP crystal**

## 4.4 FRONTIER MOLECULAR ORBITAL (FMO'S) ANALYSIS

The most important orbital in molecules are the frontier molecular orbital, called Highest Occupied Molecular Orbital (HOMO) and Lowest Unoccupied Molecular Orbital (LUMO). The HOMO represents the ability to donate an electron and LUMO as an electron acceptor, represents the ability to obtain an electron. The HOMO – LUMO analysis was performed in order to understand the electronic transitions and also to estimate the energy gap ($E_g$).

The molecular structure of DPGN and DPGP was optimized using B3LYP, hybrid function consisting of Becke's three-parameter non-local exchange function with the correlation function of Lee, Yang and Parr employing B3LYP/6-31G (d, p) basis set. However for DPGTM the same level of theory was employed with B3LYP/6-31G (d) as basis set. The HOMO – LUMO analysis was carried out for DPGN, DPGTM and DPGP from their corresponding optimized molecular structure.

The energy of the HOMO is directly related to the ionization potential, LUMO energy is directly related to the electron affinity. These orbitals determine the way in which the molecule interacts with other species. The energy gap between HOMO and LUMO characterizes the molecular chemical reactivity, optical polarizability, kinetic stability, bio-activity and chemical hardness of a molecule. As a measure of electron conductivity, this energy is a critical parameter to determine the molecular electrical transport properties. It is a well-known fact that the molecules with a small energy gap are known as soft molecule and those with large energy gap are termed as hard molecules. The hard molecules are highly stable and less reactive than the soft ones because it needs higher energy for its molecular excitation.

The HOMO and LUMO energies of the DPGN compound computed from the optimized geometry using B3LYP/6-31G (d,p) basis set are found to be -5.995 eV and -1.125 eV respectively.

Thus, the HOMO – LUMO energy gap of **DPGN** is given as,

HOMO energy, $E_{HOMO}$ = **-5.995 eV**

LUMO energy, $E_{LUMO}$ = **-1.125 eV**

HOMO – LUMO energy gap, $\Delta E_g = E_{LUMO} - E_{HOMO}$ = (- 1.125) – (- 5.995) **= 4.87 eV.**

The HOMO and LUMO energies of the DPGTM compound computed from the optimized geometry using B3LYP/6-31G (d) basis set are found to be -5.896 eV and -0.851 eV respectively.

Thus, the HOMO – LUMO energy gap of **DPGTM** is given as,

HOMO energy, $E_{HOMO}$ = **-5.896 eV**

LUMO energy, $E_{LUMO}$ = **-0.851 eV**

HOMO – LUMO energy gap, $\Delta E_g = E_{LUMO} - E_{HOMO}$ = (- 0.851) – (- 5.896) = **5.04 eV.**

The HOMO and LUMO energies of the DPGP compound computed from the optimized geometry using B3LYP/6-31G (d,p) basis set are found to be -5.880 eV and -0.536 eV respectively.

Thus, the HOMO – LUMO energy gap of **DPGP** is given as,

HOMO energy, $E_{HOMO}$ = **-5.880 eV**

LUMO energy, $E_{LUMO}$ = **-0.536 eV**

HOMO – LUMO energy gap, $\Delta E_g = E_{LUMO} - E_{HOMO}$ = (- 0.536) – (- 5.880) = **5.34 eV.**

Thus the HOMO–LUMO energy gap is 4.87 eV, 5.04 eV and 5.34 eV for DPGN, DPGTM and DPGP respectively, which implies that these crystals have high energy barriers to transfer electrons between the occupied and unoccupied orbitals. The discrepancy with the energy gap measured by experimental and theoretically computed energy gap are attributed to the fact that the theoretical computations are carried out in the gaseous phase but however the experimental calculations are based on the solid crystal sample and sample thickness. The effect of employing basis sets can also be attributed to these facts. The experimental and theoretical values of energy gap are listed in Table 4.2.

**Table 4.2 Experimental and Theoretical values of energy gap of title crystals**

| Crystals | Experimental energy gap (eV) | Theoretical energy gap (eV) | Variation (eV) |
|---|---|---|---|
| DPGN | 3.90 | 4.87 | 0.97 |
| DPGTM | 4.05 | 5.04 | 0.99 |
| DPGP | 4.14 | 5.34 | 1.20 |

The HOMO – LUMO plot of DPGN is depicted in figure 4.3. It can be understood from the plot that by considering the red shade as positive and green as negative phases the HOMO as an electron donor localizes mainly on the nitrate anion of the title compound and LUMO as an electron acceptor highly localizes on the guanidinium fragment and the phenyl rings thereby confirming the intramolecular charge transfer taking place within the molecule.

The HOMO – LUMO plot of DPGTM is depicted in figure 4.4. It can be understood from the plot that by considering the red shade as positive and green as negative phases the HOMO as an electron donor localizes mainly on the guanidinium fragment, water molecule and one part of phenyl ring of the title compound and LUMO as an electron acceptor highly localizes on the phenyl rings thereby confirming the intramolecular charge transfer taking place within the molecule.

The HOMO – LUMO plot of DPGP is depicted in figure 4.5. It can be understood from the plot that by considering the red shade as positive and green as negative phases both the HOMO and LUMO orbitals predominantly localize on the monosubstituted benzene ring. The HOMO localizes on one part of the phenyl ring, guanidinium fragments and on the oxygen atom of the phosphite ion. A highly delocalized HOMO indicates that the electrons can more readily move around the molecule and hence improved intramolecular charge transfer (ICT). On the other hand, the LUMO strongly localizes on the other part of the phenyl ring and also on the nitrogen atom of the guanidinium fragment to which the hydrogen atom has attached from the phosphite ion, indicating the presence of favorable atomic centers within DPGP for possible nucleophilic attacks and its bioactivity.

This clearly indicates that the electrons can more readily move around the molecule and hence improved intramolecular charge transfer (ICT) in the title compounds. This ICT facilitates the material's non-linear activity and molecular reactivity.

### 4.4.1 Chemical Hardness

The concept of "chemical hardness" was originally developed as a measure of the stability of molecules. One implication of this finding is that chemical hardness is related to the band gap of crystals, consistent with its being related to the LUMO - HOMO gap of molecules. The theory of chemical and physical hardnesses is useful because it unifies understanding of various properties and it connects the behaviors of molecules with those of liquids and solids.

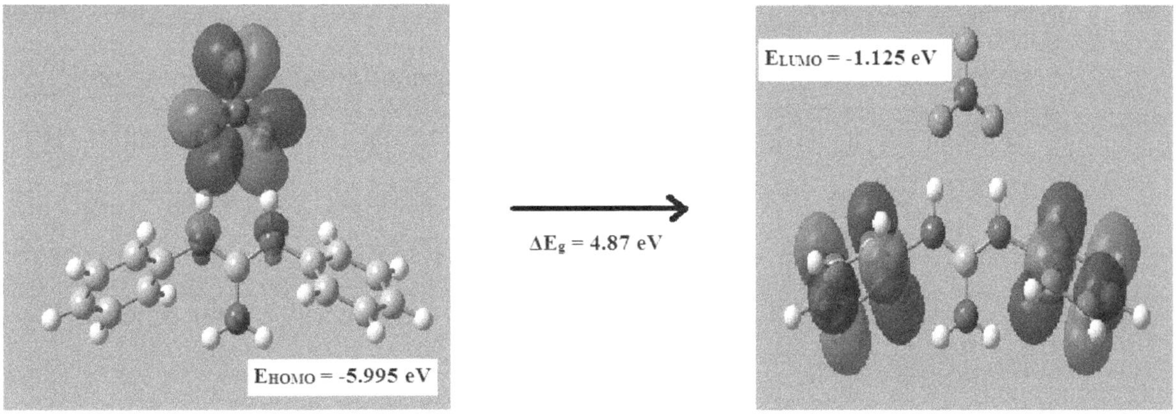


**Figure 4.3 HOMO – LUMO plot of DPGN**

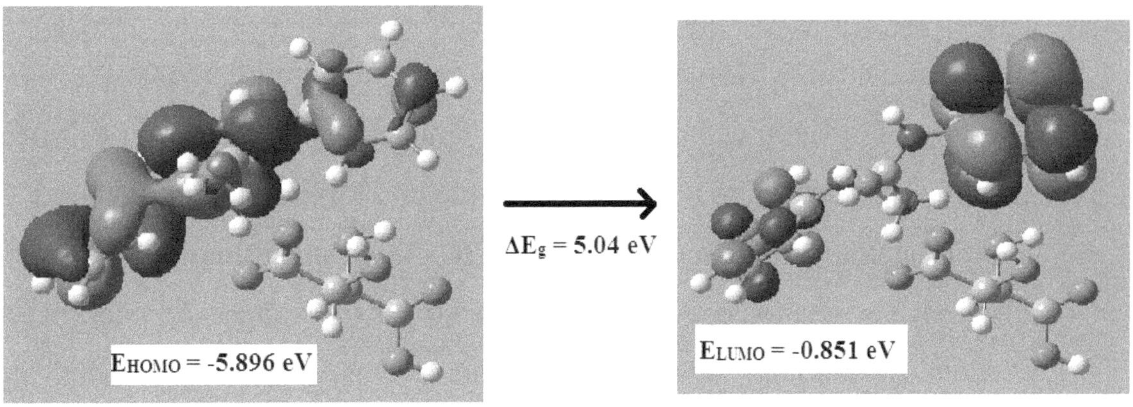


**Figure 4.4 HOMO – LUMO plot of DPGTM**

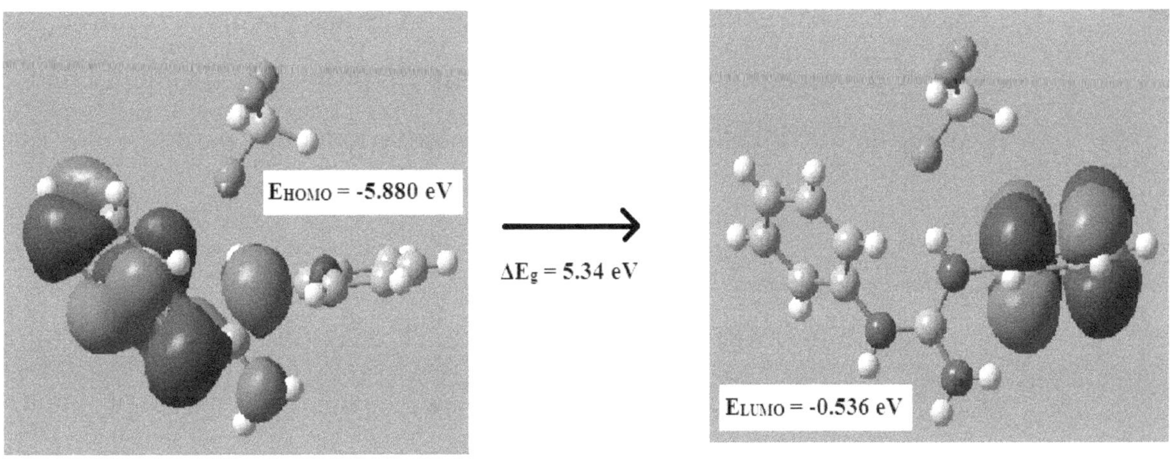


**Figure 4.5 HOMO – LUMO plot of DPGP**

Chemical hardness is an energy parameter that measures the stabilities of molecules or atoms. According to the Koopman's theorem (Koopman, 1933), the HOMO energy ($E_{HOMO}$) can be used to relate the ionization potential (IP) and the LUMO energy ($E_{LUMO}$) can be used to estimate the electron affinity (EA). If $-E_{HOMO} \approx$ IP and $-E_{LUMO} \approx$ EA, then the HOMO–LUMO energy gap is related to the chemical hardness (η) (Pearson, 1989; Pearson, 1997; Parr et. al., 1978) by the relation,

$$\eta = (IP - EA) / 2 \qquad (4.6)$$

Yang et. al., (1987) have shown that chemical and mechanical hardnesses (physical stabilities) are connected. Chemical hardness measures the resistance to a chemical change. However physical hardness measures the resistance to a physical change. Finally, it can be claimed that a material which is found to be chemically stable are expected to be physically stable too. Physical hardness can be defined to be proportional and sometimes equal to the chemical hardness (Parr and Yang, 1989). The relationship between the two types of hardness depends on the type of chemical bonding. However, for crystals, it is also found that the indentation hardness is monotonic functions of the chemical hardness density.

The high value of chemical hardness, **2.43 eV**, **2.52 eV** and **2.67 eV** for DPGN, DPGTM and DPGP indicates that it has less chemical reactivity and hence a higher chemical stability because of their HOMO – LUMO energy gap. Thereby, it can be claimed that the title compounds are chemically hard.

## 4.5 NONLINEAR OPTICAL STUDY

**Nonlinear optics** (NLO) is the branch of optics that describes the behavior of light in nonlinear media, that is, media in which the dielectric polarization **P** responds nonlinearly to the electric field **E** of the light. Nonlinear optics, a frontier in science and technology, plays a major role in the emerging era of photonics which involves photons for applications. Nonlinear optics is the study of interaction of intense electromagnetic field with materials to produce modified fields that are different from the input field in phase, frequency or amplitude. Typically, the very high optical fields from laser light produce perturbations in the optical properties of a system. Nonlinear interactions of this nature are referred to as either second or third order responses.

**Second harmonic generation** (also called **frequency doubling** or abbreviated as **SHG**) is a nonlinear optical process, in which photons with the same frequency interacting with a nonlinear material are effectively "combined" to generate new photons with twice the energy, and therefore twice the frequency and half the wavelength of the initial photons. Second harmonic generation, as an even-order nonlinear optical effect, is only allowed in mediums without inversion symmetry. This process occurs within a nonlinear medium, usually in crystals. A variety of optical combinations can be produced in most materials, which provide a host of nonlinear phenomena for applications in both basic and technological research.

Kurtz and Perry (1968) proposed a powder SHG method for comprehensive analysis of the second-order nonlinear optical effects. This is a useful method for characterizing the materials before fabrication. The limitation of this method is that the strength of the second harmonic signal is affected by the size and shape of the particle. Using this technique, Kurtz analyzed a very large number of compounds for potential applications.

### 4.5.1 Kurtz-Perry Powder Technique – Experimental Set-up

Kurtz-Perry powder technique allows one to determine the SHG nonlinearity of polycrystalline samples. The powder SHG technique requires very little instrumentation and additionally, all the instrumentation is commercially available. The conversion nonlinearity of grown crystal was determined by the modified version of Kurtz-Perry powder technique. Once the SHG of a polycrystalline sample with a few micrometer sizes has been measured, it is very straightforward to determine the SHG nonlinearity of any new material.

The crystalline sample is powdered to particle size in the range 125 – 150 μm. Relevant comparison was made with known SHG material, such as potassium dihydrogen orthophosphate (KDP) which is also grounded and sieved into the same particle size range. The powdered samples were filled separately in airtight micro-capillary tubes of uniform bore of about 1.5 mm diameter. A high intensity Nd:YAG (DCR11) laser was used as a light source. A fundamental laser beam of 1064 nm wavelength and 10 ns pulse width with 10 Hz pulse rate was made to fall normally on the sample cell. The power of the incident beam was measured using a power meter. A photomultiplier tube (PMT – Hamamatsu – model R 2059) connected to an oscilloscope (TDS 3052 B 500 MHz, Phosphor digital oscilloscope) is also required. Using the PMT and oscilloscope set-up the user can collect the SHG data and compare the results with the reference materials. A typical set-up for powder SHG measurement used in the present investigation is shown schematically in figure 4.6. The transmitted fundamental wave was passed over a monochromator (Czerny turner monochromator), which separates 532 nm (second harmonic signal) from 1064 nm, and absorbed by a $CuSO_4$ solution (F1) that removes the 1064 nm light. F2 is a BG-38 filter, which also removes the residual 1064 nm light. F3 is an interference filter with bandwidth of 4 nm and central wavelength of 532 nm. The green light was detected by a photomultiplier tube and displayed on a storage oscilloscope.

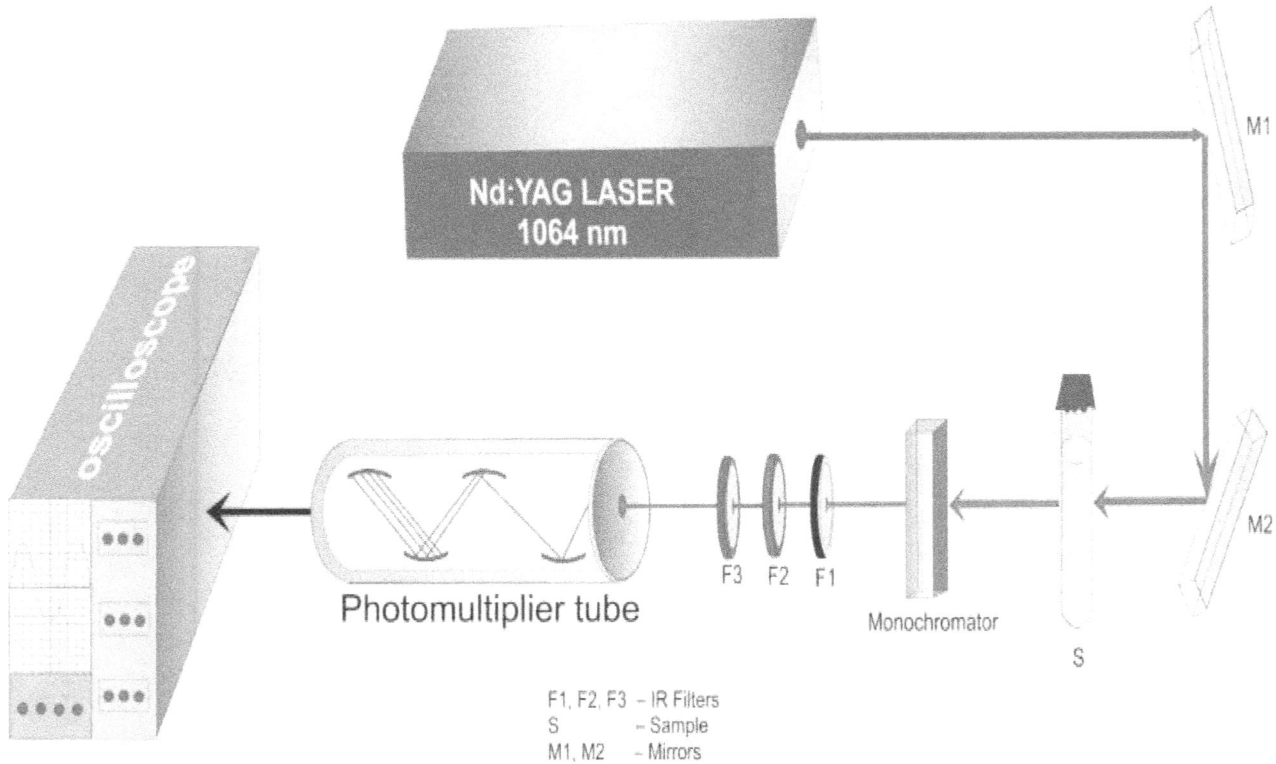


**Figure 4.6 Schematic diagram of SHG experimental set up**

### 4.5.2 SHG Nonlinearity Measurement of N,N'-Diphenylguanidine Based NLO Crystals

The second harmonic generation nonlinearity of the as-grown N,N'-diphenylguanidine based crystals in the present work was studied by Kurtz powder test. A fundamental beam from Q-switched Nd: YAG laser was passed through the powder sample of grown crystals. Enough care has been taken to maintain the same particle size for the standard material, potassium dihydrogen phosphate (KDP). The SHG behavior is confirmed from the output of the laser beam which had bright green emission ($\lambda = 532$ nm) from the powder sample.

The second harmonic signals of 14.2 mV, 14 mV and 15.2 mV were obtained for DPGN, DPGTM and DPGP powder crystalline samples respectively, for an input energy of 5 mJ/pulse, while the standard potassium dihydrogen phosphate (KDP) powder crystal sample gave a SHG signal of 15.4 mV for the same input energy. It shows that the SHG effective nonlinearities of DPGN, DPGTM and DPGP are 0.92, 0.90 and 0.98 times respectively that of standard NLO material KDP. All the grown crystals are found to be NLO active and SHG nonlinearity is almost equivalent to that of KDP. The second harmonic generation efficiencies of the grown N,N'-diphenylguanidine series of crystals are listed in Table 4.3.

**Table 4.3 Relative SHG nonlinearity of the grown crystals with respect to KDP**

| Crystal | Relative SHG nonlinearity |
|---|---|
| KDP | 1 |
| DPGN | 0.92 |
| DPGTM | 0.90 |
| DPGP | 0.98 |

### 4.5.3 Nonlinear Optical Calculations

The quantum chemical calculations are precious and inexpensive tool for predicting the molecular NLO properties of the material, such as dipole moments, polarizability and hyperpolarizability. In a medium, the nonlinear optical effects arise when the medium interact with the electromagnetic field of intense laser beam. For an isolated molecule this nonlinear response can be represented by a Taylor series expansion of the dipole moment $\mu_{tot}$ induced by the electric field (Nabil et. al., 2014, Sajan et. al., 2006).

$$\mu_{tot} = \mu_i + \sum_j \alpha_{ij} E_i + \sum_{jk} \beta_{ijk} E_i E_j \quad (4.7)$$

where, '$\mu_i$' the static dipole moment, '$\alpha$' the linear polarizability and '$\beta$' the first order hyperpolarizability. Here *i, j, k* denotes either of the *x, y, z* axes which are orthogonal.

The total static dipole moment is given by,

$$\mu_i = \left(\mu_x^2 + \mu_y^2 + \mu_z^2\right)^{1/2} \quad (4.8)$$

where $\mu_x, \mu_y, \mu_z$ are dipole moments along *x, y* and *z* directions respectively.

The static dipole moment is an important electronic property of a molecule which results from non-uniform distribution of charges on the various atoms in a molecule and is used mainly to study the intermolecular interactions involving the van der Waals type dipole–dipole forces because higher the dipole moment, stronger will be the intermolecular interactions.

Polarizability (α) is the ability for a molecule to be polarized which is a property of matter. Polarizability determine the dynamical response of a bound system to external fields and provide insight into the internal structure of a molecule.

The mean polarizability is given as,

$$\alpha_{tot} = \frac{\alpha_{xx} + \alpha_{yy} + \alpha_{zz}}{3} \tag{4.9}$$

where, $\alpha_{xx}, \alpha_{yy}, \alpha_{zz}$ the diagonal components of polarizability tensor.

The first hyperpolarizability ($\beta$) is a third rank tensor that can be described by 3×3×3 matrix. The 27 components of 3D matrix can be reduced to 10 components due to Kleinman symmetry (Kleinman, 1962). The output from Gaussian 03 provides 10 components of this matrix as,

$$\beta_{xxx}, \beta_{xxy}, \beta_{xyy}, \beta_{yyy}, \beta_{zxx}, \beta_{xyz}, \beta_{zyy}, \beta_{xzz}, \beta_{yzz}, \beta_{zzz} \tag{4.10}$$

The components of first hyperpolarizability can be calculated using the following equation,

$$\beta_i = \beta_{iii} + \frac{1}{3}\sum_{i \neq j} \beta_{ijj}\beta_{jij}\beta_{jji} \tag{4.11}$$

The complete equation for calculating the magnitude of first hyperpolarizability from the *x, y* and *z* components obtained from GAUSSIAN 03 output is given as,

$$\beta_{tot} = (\beta_x^2 + \beta_y^2 + \beta_z^2)^{1/2} \tag{4.12}$$

$$\beta_{tot} = \left[(\beta_{xxx}\beta_{xyy}\beta_{xzz})^2 + (\beta_{yyy}\beta_{yzz}\beta_{yxx})^2 + (\beta_{zzz}\beta_{zxx}\beta_{zyy})^2\right]^{1/2} \tag{4.13}$$

Since the Gaussian 03 output values of the polarizability and the hyperpolarizability tensors components are reported in atomic units (a.u.), the calculated values have been converted into electrostatic units (esu). ($\alpha$: 1 a.u. = 0.1482 × $10^{-24}$ esu ; $\beta$ : 1 a.u. = 8.3693 × $10^{-33}$ esu).

The computed values of dipole moment ($\mu_i$) and polarizability ($\alpha$) of DPGN, DPGTM and DPGP compounds with B3LYP/6-31 G (d,p), B3LYP/6-31 G (d), B3LYP/6-31 G (d,p) methods are 10.421 D, 4.3552 D, 9.9678 D respectively and 21.95 × $10^{-24}$ esu, 26.17 × $10^{-24}$ esu and 21.01 × $10^{-24}$ esu respectively.

The calculated first order hyperpolarizability, $\beta_{tot}$ of DPGN, DPGTM and DPGP was found to be 84.052$\times$ $10^{-31}$ esu, 30.2353 $\times$ $10^{-31}$esu and 27.057$\times$ $10^{-31}$ esu. This high value of ($\beta$) and the non-zero value of ($\mu_i$) are responsible for the enhancement of second harmonic generation properties of the title compounds at molecular level. Since potassium dihydrogen phosphate (KDP) is one of the prototypical molecules used in the study of the NLO properties of molecular systems, it is used frequently as a threshold value for comparative purposes. The first order hyperpolarizability of KDP computed at B3LYP/6-31G (d) level of basis set was found to be $\beta_{KDP} = 6.85 \times 10^{-31}$ esu (Elleuch et. al., 2014). Thus, the first order hyperpolarizability of the title compounds are found to be higher with respect to KDP. The theoretical values of fundamental NLO parameters of the title compounds obtained through the finite-field approach are listed in Table 4.4.

It could be observed that DPGN, DPGTM and DPGP are polar compounds having non-zero dipole moment and polarizability values which provide microscopic origin of its NLO behavior for these materials. The experimental results of SHG can be compared with that of the predicted first hyperpolarizability. The larger value of predicted hyperpolarizability than that of experimental value may be validated, since the theoretical value belongs to the isolated molecule in the gas phase whereas the latter one was attributed to the bonded molecule in the solid state (Sudharsana et. al., 2012). Additionally, the solid-state structure configuration is associated with several intermolecular forces, such as van der Waals interactions and crystal field interactions. Moreover, the experimental parameters, such as particle size dependence and input energy of laser can also be validated for the variation in the predicted and the experimental nonlinear effects of the title compounds.

**Table 4.4 Theoretical values of fundamental NLO parameters of the title compounds**

| PARAMETERS | DPGN | DPGTM | DPGP |
|---|---|---|---|
| Dipole moment (Debye) | | | |
| $\mu_x$ | 0.0570 | -1.5786 | 0.9584 |
| $\mu_y$ | -10.415 | 3.9203 | -9.8409 |
| $\mu_z$ | 0.3503 | -1.0520 | 1.2630 |
| $\boldsymbol{\mu_i}$ | **10.421** | **4.3552** | **9.9678** |
| Polarizability (a.u.) | | | |
| $\alpha_{xx}$ | 281.144 | 271.8553 | 251.8438 |
| $\alpha_{yy}$ | -0.1218 | 39.7992 | 9.0547 |
| $\alpha_{zz}$ | 163.414 | 218.2618 | 164.4810 |
| $\boldsymbol{\alpha_{tot}\ (esu)}$ | **21.95 × $\mathbf{10^{-24}}$** | **26.1778 × $\mathbf{10^{-24}}$** | **21.01 × $\mathbf{10^{-24}}$** |
| Hyperpolarizability (a.u.) | | | |
| $\beta_{xxx}$ | 26.6281 | 223.1613 | 372.4884 |
| $\beta_{xxy}$ | -598.0263 | -34.0459 | 223.4325 |
| $\beta_{xyy}$ | 3.7815 | -113.6956 | -60.1140 |
| $\beta_{yyy}$ | -393.6515 | 18.0123 | -32.7303 |
| $\beta_{zxx}$ | 1.5686 | -178.7613 | 51.2286 |
| $\beta_{xyz}$ | -73.9521 | -36.8916 | -91.2095 |
| $\beta_{zyy}$ | 4.5471 | -96.3606 | -36.5011 |
| $\beta_{xzz}$ | 0.1631 | -62.0730 | -43.9800 |
| $\beta_{yzz}$ | -12.1439 | 4.6207 | -16.9713 |
| $\beta_{zzz}$ | -1.1512 | -82.8395 | 33.2247 |
| $\boldsymbol{\beta_{tot}\ (esu)}$ | **84.05 × $\mathbf{10^{-31}}$** | **30.23× $\mathbf{10^{-31}}$** | **27.05× $\mathbf{10^{-31}}$** |

## 4.6 LASER DAMAGE THRESHOLD CHARACTERISTICS

The system based on nonlinear optical technology requires high peak powers to maximize frequency conversion. The operation of nonlinear optical devices obviously involves the exposure of material to high power laser source. For harmonic generation, the efficiency of conversion is strongly dependent on the incident power level. The utility of NLO crystal depends not only on the linear and nonlinear optical properties but also largely on its ability to withstand high power lasers. Hence, the subject of Laser Damage Threshold (LDT) is of great importance to the design and successful operation of nonlinear devices.

The bulk materials appear more damage-resistance than the surface. Since, the performance of the crystal mainly depends on surface quality, the study of surface damage due to incident laser light is important (Gupta et. al., 2002; Krol et. al., 2005). The mechanism of laser induced damage in materials is often very complex involving various processes, like electron avalanche, multi photon absorption, photo ionization by thermal absorption and photo chemical dislocation etc., (Vijayan et. al., 2007).

From the earlier report (Stuart et. al., 1995, Vivek and Murugakoothan, 2013), it is known that in the long-pulse regime ($\tau > 100$ ps), the damage is controlled by the rate of thermal conduction through the atomic lattice and in the short-pulse ($\tau < 100$ ps), it is due to optical break down and various nonlinear ionization mechanism. Hence, thermal effect becomes important for the pulse width that are nanosecond and longer. The local heating caused by macroscopic inclusion is also a more severe problem in the high intensity short-pulse system than in low-intensity long-pulse. In the present work we have used 10 ns pulse.

However, the LDT of wide band gap materials depends on

(i). the pulse duration,

(ii). pulse frequency,

(iii). incubation effects,

(iv). impact beam diameter and

(v). the band gap.

### 4.6.1 Instrumentation

Optical damage may be caused by several mechanisms and can be of various types. The predominant type is irreversible damage, which can take the form of surface or bulk damage. It can be caused by thermal heating, induced absorption, self-focusing and dielectric breakdown. When performing LDT testing, the sample is irradiated numerous times using a small beam over the whole clear aperture of the sample (Boling et. al., 1973).

The schematic diagram of the LDT setup is shown in figure 4.7. The laser damage threshold value was measured using a Q-switched Nd:YAG laser source of pulse width 10 ns and repetition rate of 10 Hz operating in $TEM_{00}$ mode. The energy per pulse of 1064 nm laser radiation attenuated using appropriate neutral density filters was measured using an energy meter (Coherent EPM 200). For both single and multiple shots experiments, the sample was mounted on an X–Y translator which facilitates in bringing different areas of the sample. Each shot has to be made at different places of the crystal to avoid the cumulative effect. For surface damage observation, the sample was placed at the focus of a plano-convex lens with focal length of 20 cm. The laser beam of diameter 2 mm was focused on the crystal sample. The pulse energy of each shot was measured using the combination of a phototube and an oscilloscope. During laser irradiation, the power meter records the energy density of the input laser beam by which the crystal gets damaged.

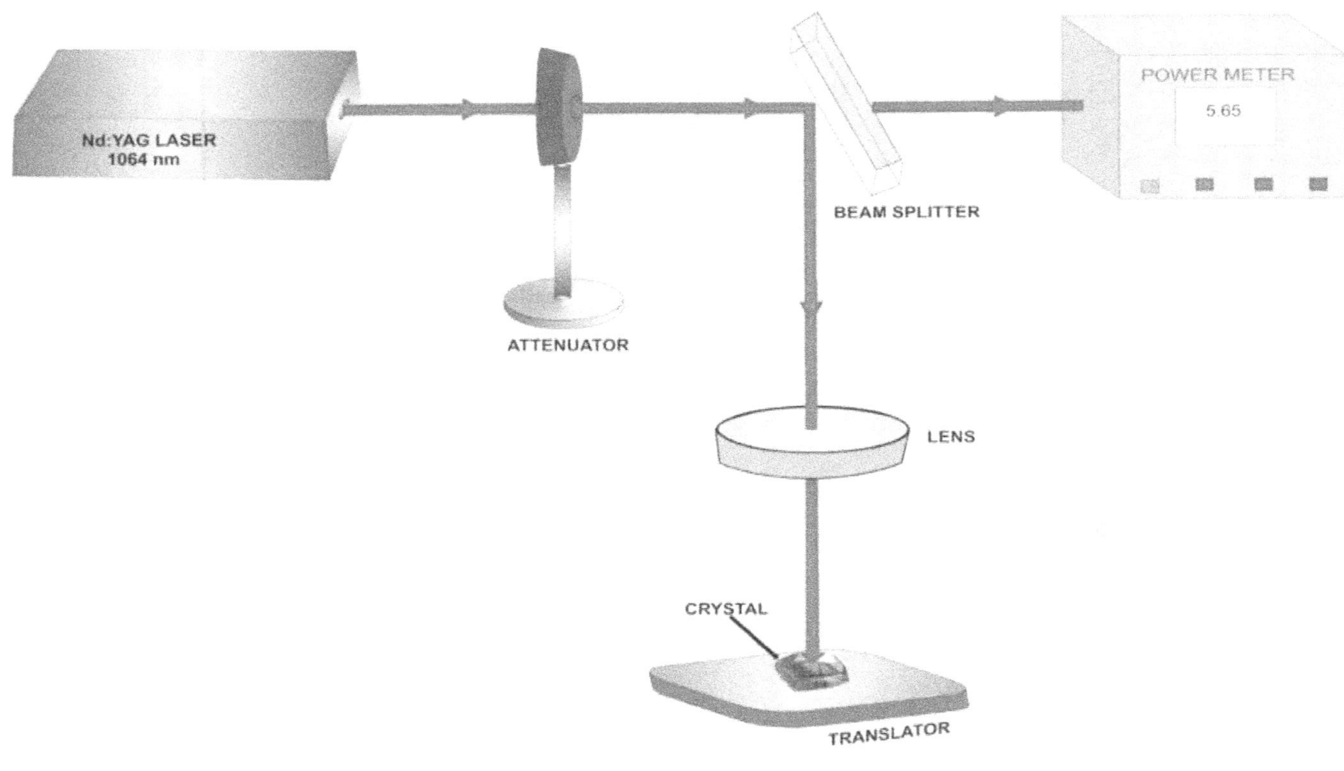


**Figure 4.7 Schematic diagram of laser damage threshold setup**

### 4.6.2 LDT Measurement of N,N'-Diphenylguanidine Based Single Crystals

In the present study, an actively Q-switched Nd:YAG laser in $TEM_{00}$ mode with pulse width 10 ns and repetition rate 10 Hz was used for the laser induced damage threshold study. The radius of laser beam was 1 mm. A well-polished sample of the grown crystals with clean surface and prominent face [DPGN – (010), DPGTM – (100) and DPGP – (110)] was subjected to the LDT study. The sample was mounted on the goniometer used to align the sample at the desired position. The energy density of the input laser beam for which the crystal gets damaged was recorded using a power meter.

Single and multiple shot laser damage measurements were made on the polished surface of prominent plane of the grown crystals. The energy density is calculated using the formula (Anandha Babu et. al., 2010; Prabha et. al., 2010)

$$\text{Energy density} = E / A \qquad (\text{J cm}^{-2}) \qquad (4.14)$$

where, E the input energy (mJ) and A the area of the circular spot (mm).

Hence, the power density is calculated using the expression,

$$\text{Power density} = E / \tau A \qquad (\text{GW cm}^{-2}) \qquad (4.15)$$

where, τ the pulse width (ns).

Energy density or energy fluence is a measure used to describe the energy delivered per unit area. The energy fluence of the title compounds measured for both single and multiple shot methods are given in Table 4.5.

The laser damage threshold value was found to be lowered when the crystals was subjected to multiple shot laser damage. The LDT experiments of single shot was carried out at 1 pulse per second and multiple shot was carried out at 10 pulses per second. It shows that single shot laser damage is higher than the multiple shot laser damage.

**Table 4.5 Energy fluence for the grown crystals**

| Crystals | Single shot | | Multiple shot | |
|---|---|---|---|---|
| | **Input energy (mJ)** | **Energy fluence ($J\ cm^{-2}$)** | **Input energy (mJ)** | **Energy fluence ($J\ cm^{-2}$)** |
| DPGN | 8.7 | 28.0 | 6.5 | 20.7 |
| DPGTM | 4.2 | 13.5 | 4.1 | 13.1 |
| DPGP | 2.0 | 6.5 | 1.8 | 6.0 |

The single shot measurements of the title compounds in comparison with the LDT data for some other standard NLO crystals are given in Table 4.6. From these data it is clear that the title compounds show noticeable values of LDT.

**Table 4.6 LDT data for the reference and grown crystals**

| Crystals | Single shot LDT ($GW\ cm^{-2}$) | Multiple shot LDT ($GW\ cm^{-2}$) |
|---|---|---|
| KDP | 0.20* | - |
| UREA | 1.50* | - |
| DPGN | 2.80 | 2.07 |
| DPGTM | 1.34 | 1.31 |
| DPGP | 0.65 | 0.60 |

*(Vijayan et. al., 2007)

In addition to the thermal effect, multi-photon ionization is the important cause of laser induced damage. For pulse widths that stretch into several nanoseconds, thermal effects are unavoidable while for picosecond pulse widths thermal effects are negligible. This may be due to the fact that thermal effects take several nanoseconds to build up and could take several milliseconds to decay. Thus it can be claimed that the laser damage of the title compounds are of thermal origin.

## 4.7 CONCLUSION

The linear and nonlinear optical responses of DPGN, DPGTM and DPGP crystals were discussed in this chapter. The linear optical property of the title compounds were assessed by UV-vis spectral analysis. The transmittance spectra show that the grown crystals have good transparency window in the UV-vis region which is an important parameter for fabrication of nonlinear optical devices. The band gap of the title compounds were determined from the transmittance spectrum. It is also observed that the experimental band gap of the grown crystals are in the range between 3.90 – 4.14 eV, thereby favoring optical applications. The HOMO – LUMO band gap for the title compounds were computed and found to be comparable with the experimental ones. From HOMO – LUMO energy gap, the chemical hardness of the title compounds were calculated and found to chemically stable and thereby paving way for physical hardness. The existence of second harmonic generation in the title compounds were assessed by the Kurtz-Perry powder technique. From the experimental results it can be concluded that all the title compounds are NLO active and their nonlinearity is almost equivalent to that of KDP. The nonlinear optical parameters, such as dipole moment, polarizability and first order hyperpolarizability were computed in order to understand NLO behaviour at molecular level. The nonlinear optical parameters of title compounds were found to be non-zero and greater than that of KDP. The laser damage threshold study was carried out for the title compounds in order to estimate their stability towards laser. The LDT study confirms that the grown crystals exhibit higher stability to laser and high value of LDT compared with KDP. The promising linear and nonlinear optical properties suggest that the grown N,N'-diphenylguanidine based NLO crystals are suitable for fabricating optical element for nonlinear optical applications.

# CHAPTER – V

# ELECTRICAL, THERMAL AND MECHANICAL PROPERTIES OF N,N'-DIPHENYLGUANIDINE BASED NLO SINGLE CRYSTALS

## 5.1 INTRODUCTION

In order to realize the use of nonlinear optical (NLO) crystals for device fabrication, they have to be investigated by electrical, thermal and mechanical techniques. Dielectric studies are used to analyze the types of polarization present in the material and also reveal the quality of the crystal. Thermo-Gravimetric (TG) and the Differential Thermal Analyses (DTA) are two very important analyses to throw light on the thermal stability of the materials. Thermo-Gravimetric Analysis can be used to understand the chemistry of decomposition of a particular compound and also provides information about the temperature range over which a particular sample appears to be stable or unstable. Differential Thermal Analysis (DTA) is a thermo-analytic technique. A DTA thermogram provides data on the transformations that have occurred, such as glass transition, crystallization, melting and sublimation. Hardness is a measure of material resistance to localized plastic deformation. The hardness of the material depends on different parameters, such as lattice energy, Debye temperature and inter-atomic spacing. The knowledge about the electrical behavior, thermal and mechanical stability play a key role in the fabrication of NLO devices. In the present study, the family of N,N'-diphenylguanidine based novel potential NLO crystals grown was subjected to dielectric studies, TG-DTA analyses and Vicker's micro hardness test to reveal their electrical, thermal and mechanical properties.

## 5.2 DIELECTRIC STUDIES

When an external electric field is applied to an ionic solid, positive charges are displaced in the direction of field towards negative end whereas negative charges are displaced in the opposite direction. This displacement produces dipoles throughout the solid. The dipole moment per unit volume of the solid is the vector sum of all individual dipole moments within the volume and is called the polarization **P** of the solid.

The polarization occurs due to several mechanisms. Electronic polarization is the result of displacement of positively charged nucleus and electrons of an atom in opposite directions by the application of electric field. The electron cloud around the nucleus shifts towards positive end of the field when the field is applied. Such a shift results in a dipole moment within the atom equal to the product of charge and shift distance and proportional to the field strength. The constant of proportionality is called electronic polarizability of the atom, which is independent of temperature.

During the chemical bonding, atoms may acquire excess negative or positive charge and form an ionic bond. When an electric field is applied to ionic solid, cations and anions get displaced in opposite directions. Ionic polarization is also independent of temperature. In some molecule, center of positive and negative charges do not coincide and give rise to orientational polarization. Even in the absence of electric field orientational polarization exists. This polarization depends on temperature. The fourth type of polarization called space charge polarization occurs due to accumulation of charges at the electrodes or at the interfaces in the multiphase material.

Total polarization is the sum of the contributions from various compounds,

$$\boldsymbol{P}_{total} = \boldsymbol{P}_e + \boldsymbol{P}_i + \boldsymbol{P}_o + \boldsymbol{P}_s \qquad (5.1)$$

Electronic polarization is the fastest and it will complete at the instant of the applied field. At high optical frequencies electronic polarization occurs. Ionic polarization is slower than the electronic polarization. This is because, ions are heavier than electron cloud. The ionic polarization occurs in the infrared frequency range. Hence at optical frequencies, there is no ionic polarization. Orientational polarization is slower than ionic polarization and it occurs when a frequency of the applied voltage is in the audio range. Space charge polarization is the slowest process. The decrease in the value of dielectric constant with the frequency takes place when the jumping frequency of electric charge carriers cannot follow the alternation of the ac electric field applied beyond a certain critical frequency (Ponpandian et. al., 2002). In orientational polarization, the randomizing action of thermal energy decreases the tendency of permanent dipoles to align themselves in the applied field. This results in decrease in the dielectric constant with increasing temperature. The other effect of temperature is to facilitate the diffusion of ions in space charge polarization.

### 5.2.1 Theory

In an anisotropic material polarization related to dielectric polarization **P** and electric field **E** is related to electric displacement vector **D** as

$$\boldsymbol{D} = \varepsilon \boldsymbol{E} + \boldsymbol{P} \qquad (5.2)$$

where $\varepsilon$ the permittivity of the medium. For a parallel plate capacitor with a dielectric in between the plates, the capacitance C related to area and spacing d is given as:

$$C = \frac{\varepsilon A}{d} = \frac{\varepsilon_0 \varepsilon_r A}{d} \qquad (5.3)$$

where $\varepsilon_r$ the dielectric constant of the material and it depends on external factors, such as frequency, voltage, temperature, pressure and humidity.

### 5.2.2 Experimental

The dielectric constant and the dielectric loss of the grown crystals were studied at different temperature using HIOKI 3532 LCR HITESTER in the frequency region from 100 Hz to 5 MHz. Sample of uniform cross-sectional area cut along the growth axis was polished and placed between the two copper electrodes, which acts as a parallel plate capacitor. Silver paint was coated on the surface of the sample in order to make firm electrical contact.

### 5.2.3 Dielectric Study of N,N'-Diphenylguanidine Based NLO Single Crystals

The capacitance and dielectric loss of the samples were measured for various frequencies, in the range 100 Hz – 5 M Hz at different temperatures. The dielectric constant is calculated using the relation, $\varepsilon_r = Ct/\varepsilon_0 A$, where C the capacitance, t the thickness of the sample, $\varepsilon_0$ the permittivity of free space and A the area of cross section of the sample. The observations were made while cooling the sample. The plot of dielectric constant ($\varepsilon_r$) and dielectric loss (tan δ) versus applied frequency for N,N'-diphenylguanidine based crystals are shown in figures 5.1 and 5.2 respectively.

Typical sample dimensions of DPGN, DPGTM and DPGP crystals were $10.72 \times 7.15$ mm$^2$, $9.51 \times 4.83$ mm$^2$, $8.68 \times 4.42$ mm$^2$ in surface area and 2.04 mm, 1.58 mm, 2.64 mm in thickness. The dielectric constant measurements for DPGN, DPGTM and DPGP crystals were carried out on their prominent (010), (100) and (110) planes respectively. The dielectric constant has high values in the low frequency region and then it decreases with further increase of applied frequency. The dielectric constant decreases very rapidly at low frequencies and slowly at higher frequencies. The electronic exchange between the ions in the crystal causes local displacement of electrons in the direction of the applied field that causes polarization. This is a normal dielectric behaviour that both dielectric constant and dielectric loss decrease with an increase in frequency. The very high value of $\varepsilon_r$ at lower frequencies may be attributed to the presence of all four polarizations, namely space charge, orientational, electronic and ionic polarizations and its low value at higher frequencies may be due to the loss of significance of these polarizations gradually.

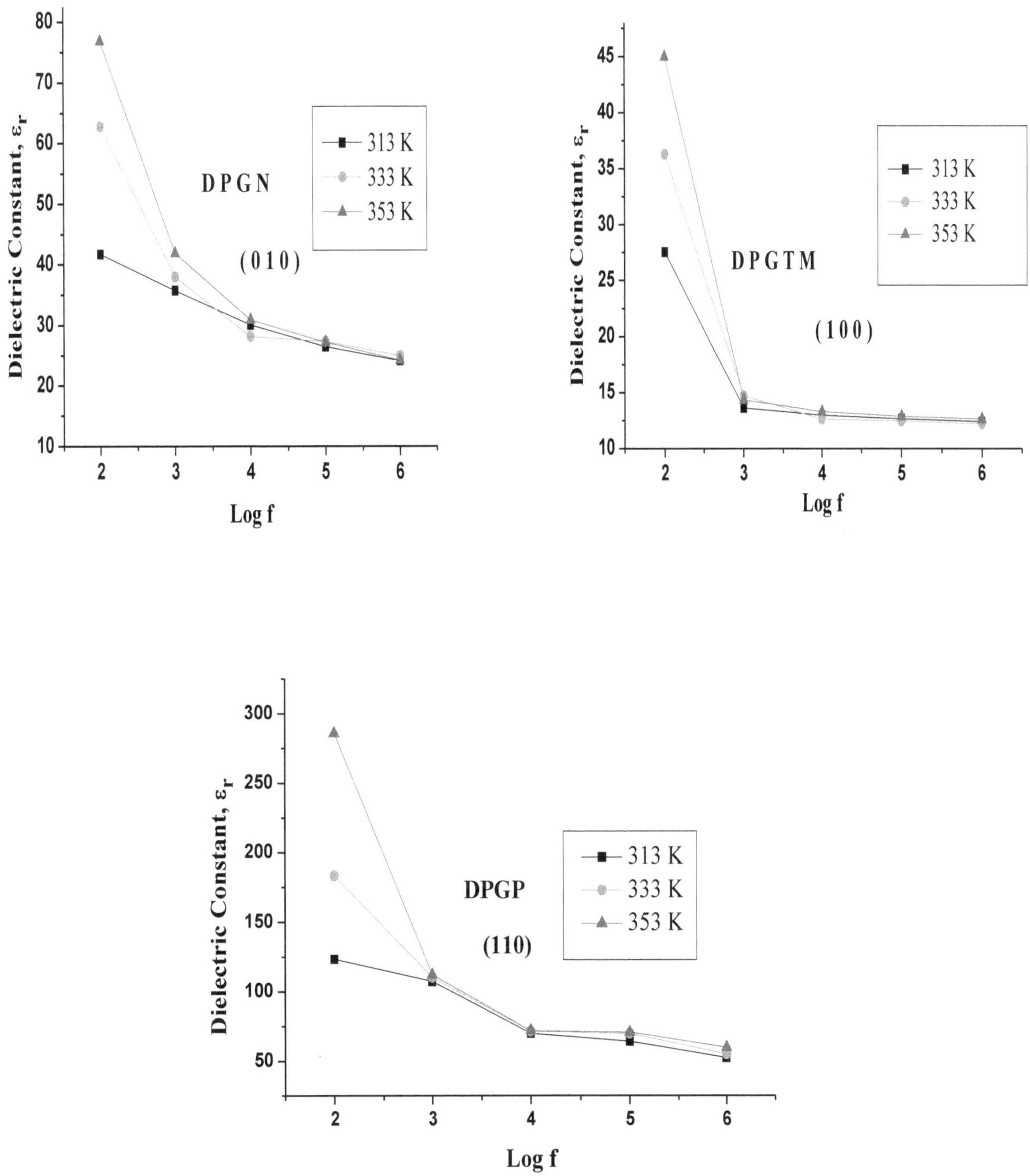


**Figure 5.1 Frequency response of dielectric constant ($\varepsilon_r$) for variation in temperature of DPGN, DPGTM and DPGP crystals**

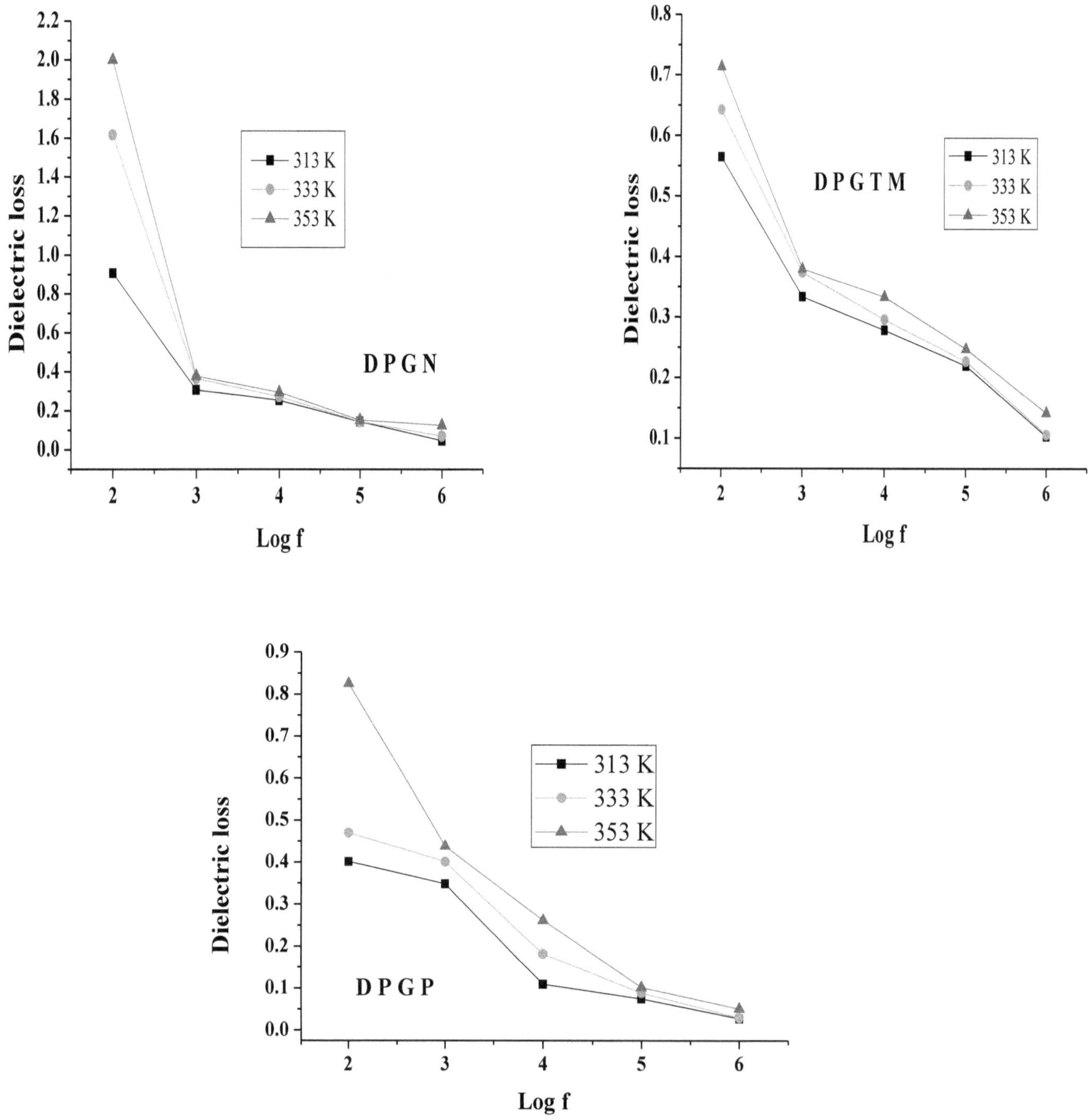


**Figure 5.2 Frequency response of dielectric loss for variation in temperature of DPGN, DPGTM and DPGP crystals**

The space charge contribution will depend on the purity and perfection of the material and it has noticeable influence in the low frequency region. The higher values of dielectric constant at lower frequencies may be attributed to space charge polarization and also to crystal lattice defects. As the frequency increases, a point is reached, where the space charge cannot sustain and comply with the external field and hence polarization decreases with frequency giving rise to diminishing value of $\varepsilon_r$ and dielectric loss (tan δ). Continuous gradual decrease in dielectric loss (tan δ) as well as $\varepsilon_r$ suggests that the crystals, like any normal dielectrics, may have domains of different sizes and varying relaxation times.

In accordance with Miller rule, the lower value of dielectric constant is a suitable parameter for the enhancement of SHG efficiency (Debrus et. al., 2002). The low value of dielectric loss (tan δ) in the region of high frequency for the sample suggests that the grown crystals possess enhanced optical quality with lesser number of electrically active defects and this parameter is of vital importance for non-linear optical materials in their applications (Rak et. al., 2005).

## 5.3 THERMAL ANALYSES

Recent trends indicate that the thermal analysis has become an established method in the study of the thermal behaviour of materials and finds widespread applications in diverse industrial and research fields. Thermal analysis is a general term, which covers a group of related techniques in which the temperature dependent parameters of any physical property of substance is measured. In addition to provide valuable information on the thermal stability of the compounds thermal analysis provides a detailed knowledge about the decomposition of products, adsorbed moisture content and solvent residues of materials. The importance of thermal analysis in quality control, failure analysis and material research and development is well established.

### 5.3.1 Thermoanalytic Method

Thermo analytical methods involve the measurement of various properties of materials subjected to dynamically changing environments under predetermined condition of heating rate, temperature range and gaseous atmosphere or vacuum. In many cases, the use of single thermo analytical technique may not provide sufficient information to solve the problem on hand and hence the use of other thermal technique, either independently or simultaneously for complementary information, becomes necessary. For example, both differential thermal analysis (DTA) and thermogravemetric (TG) are widely used in studies involving physicochemical changes accompanied by variation in the heat content and the weight of the material. Among the thermal methods, the most widely used technique are TG, DTA and DSC which find the extensive use in all fields of inorganic chemistry, organic chemistry, metallurgy, mineralogy and many other areas.

### 5.3.2 Thermogravimetry

Thermogravimetry is a technique in which the mass of a substance is measured as a function of temperature, while the substance is subjected to a controlled temperature program. The curves obtained in the thermogravimetric analysis are called a thermogram (TG) and its first derivative is called a derivative or differential thermogram (DTG). The inflection point in the thermogram corresponds to the peak point in the derivative thermogram.

### 5.3.3 Instrumentation

The modern commercial thermogravitometric analysis instrument consists of

(i). A sensitive analytical balance,

(ii). A temperature programmable furnace,

(iii). A pure gas system for providing suitable gas atmosphere,

(iv). A microprocessor for instrument control, data acquisition and display.

Even though different types of balance mechanism are available today, those employing null-point weighing mechanism is favored as the sample remains in the same zone of the furnace irrespective of changes in mass. The furnace is normally an electrical resistive heater and the temperature range for most of the furnace is from ambient to 1000–2000 °C. Thermogravimetry is widely used to determine the thermal stability, decomposition temperature, oxidative stability etc.

The rate of heating affects the results of thermal analysis. The rate of heat exchange between the furnace and the sample depends upon the heating rate which influences the TG curve in a number of ways. A slower rate gives a better resolution of the closely lying steps, while the faster heating rate merges such steps. The type and material of the sample holder affect the TG curve to a considerable extent. The shape and size of the sample holder determine the thermal gradients within the sample and the area of sample–environmental gas interface. One of the objectives of TG and DTA is to delineate as accurately as possible the various temperatures associated with the thermal behaviour of a given substance, i.e., temperature at which reaction is complete. Often, the values reported from TG measurements differ considerably from the true value. Some of the reasons for this are the effects of self-generated atmosphere and heating rate as well as the lag between the sample and furnace temperature (Douglas et. al., 2004; Usharani, 2000). Hence in the present study, the thermal characteristic of the grown DPGN, DPGTM and DPGP crystals were studied by simultaneous TG-DTA technique between 30 °C and 500 °C at a heating rate of 2 °C/min. in the nitrogen atmosphere using NETZSCH STA 449F3.

### 5.3.4 Theoretical Elemental Composition Analysis of N,N'-Diphenylguanidine Based NLO Crystals

The presence of various elements in the title compounds were computed theoretically using CHEM DRAW ULTRA chemistry software. These data are useful in interpreting the thermal stability of the title compounds. The elemental composition of title compounds are presented in Table 5.1.

**Table 5.1 Elemental composition of grown crystals**

| Crystals | Elemental composition analysis (%) |
|---|---|
| DPGN | C – 56.93%, H – 5.14%, N – 20.43%, O – 17.50% |
| DPGTM | C – 53.82%, H – 5.58%, N – 11.08%, O – 29.52% |
| DPGP | C – 53.24%, H – 5.50% N – 14.33%, O – 16.37% P – 10.56% |

### 5.3.5 Thermal analysis of N,N'-Diphenylguanidine Based NLO Crystals

The thermal stability of DPGN was studied by thermo gravimetric (TG) and Differential thermal analyses (DTA). The DPGN sample weighing 10.672 mg was analysed and the thermogram is depicted in figure 5.3. From the TG curve, it is evident that the material is stable up to 192 °C and moisture free. The DTA curve shows an endothermic peak at 198 °C, with corresponding weight loss in TG, is evident that the given thermal energy may be utilized to overcome the hydrogen bonding between the guanidinium cation and nitrate anion, which can be attributed to the decomposition of the sample. The TG curve shows two stage weight loss pattern when the material was heated from 34 to 500 °C. The first major weight loss was occurred between the temperatures 198 and 260 °C with the elimination of 40.18 % of the material into gaseous products, such as nitrogen oxides, CO, $NH_2$ and $CO_2$ as seen in the low temperature region. The second stage weight loss noticed between the temperatures 260 and 498 °C experiences a weight loss of about 27.95%. The residue left out at the end is about 31.40 % by weight at 498.1 °C. This may be due to the residual carbon mass at the end of the decomposition reactions as the theoretical elemental analysis shows that the percentage of carbon in the title compound is nearly 56.93%. From the above discussions it can be concluded that the DPGN crystals can be used for NLO applications till 192 °C.

The DPGTM sample weighing 7.712 mg was analyzed and the thermogram is depicted in figure 5.4. In TG trace, four stage weight loss pattern is observed for DPGTM. The first weight loss pattern observed between 100 and 162 °C is assigned to loss of water of crystallization with elimination of 4.42% as water vapors. The immediate second, third and fourth stages of weight losses occurred between 162 - 240°C, 240-360 °C and 360-600°C with the elimination of 46.01 % , 21.94 %, and 12.85% of the material into gaseous products, such as nitrogen oxides, $NH_2$, CO, $CO_2$ of diphenylguanidinium tartrate as seen in the low temperature region. Based on the decomposition, it is confirmed that N,N'-diphenylguanidine and tartaric acid are not separate entities but it is N,N'-diphenylguanidinium tartrate. The endotherm maximum at 162 °C corresponds to the degradation of DPGTM. Hence it is confirmed that the DPGTM is formed and it also revealed that the title compound decomposes before melting. The residue left at the end at (600° C) is about 14.78% by weight. This may be due to the residual carbon mass at the end of the decomposition reactions. Thus DPGTM could be used up to 100 °C for NLO applications.

The thermal stability of DPGP was studied by thermo gravimetric (TG) and differential thermal analyses (DTA). The DPGP sample weighing 4.359 mg was analysed and the thermogram is depicted in figure 5.5. From the TG curve, it is evident that the material is thermally stable up to 224 °C and moisture free. A major weight loss is found to occur at 224 °C. The sharpness of weight loss in thermogram ascertain the phase purity and quality of the DPGP crystal. An endotherm at 224 °C in DTA curve corresponds to weight loss in TG curve. As there is no endotherm before weight loss of DPGP compound, it is evident that DPGP decomposes before melting. Therefore, it is established that the DPGP crystal could be used to any application up to 224 °C. The residue left at the end is about

36.60 % by weight. This may be due to the residual carbon mass at the end of the decomposition reactions.

From this study it can be concluded that the compounds studied in the present work can be used for nonlinear optical applications in the working temperature region. In order to have an idea about the thermal stability of the compounds grown in the present study, the thermal stability of these compounds are compared with some of the reported compounds in Table 5.2.

**Table 5.2 Thermal stability for some reported materials**

| Materials | Thermal stability (°C) |
|---|---|
| Guanidinium carbonate [a] | 190 |
| N,N'-Diphenylguanidine [b] | 146 |
| Guanidinium nitrate [a] | 215 |
| Guanidinium L-monohydrogen tartrate [c] | 182 |
| Zinc guanidinium phosphate [d] | 227 |
| N,N'-Diphenylguanidinium nitrate (DPGN)* | 192 |
| N,N'-Diphenylguanidinium hydrogen L-tartrate monohydrate (DPGTM)* | 100 |
| N,N'-Diphenylguanidinium dihydrogen phosphite (DPGP)* | 224 |

* Present work, [a] (Wendlandt et. al., 1984), [b] (Hu et. al., 2012), [c] (Arumanayagam and Murugakoothan, 2012), [d] (Suvitha and Murugakoothan, 2012).

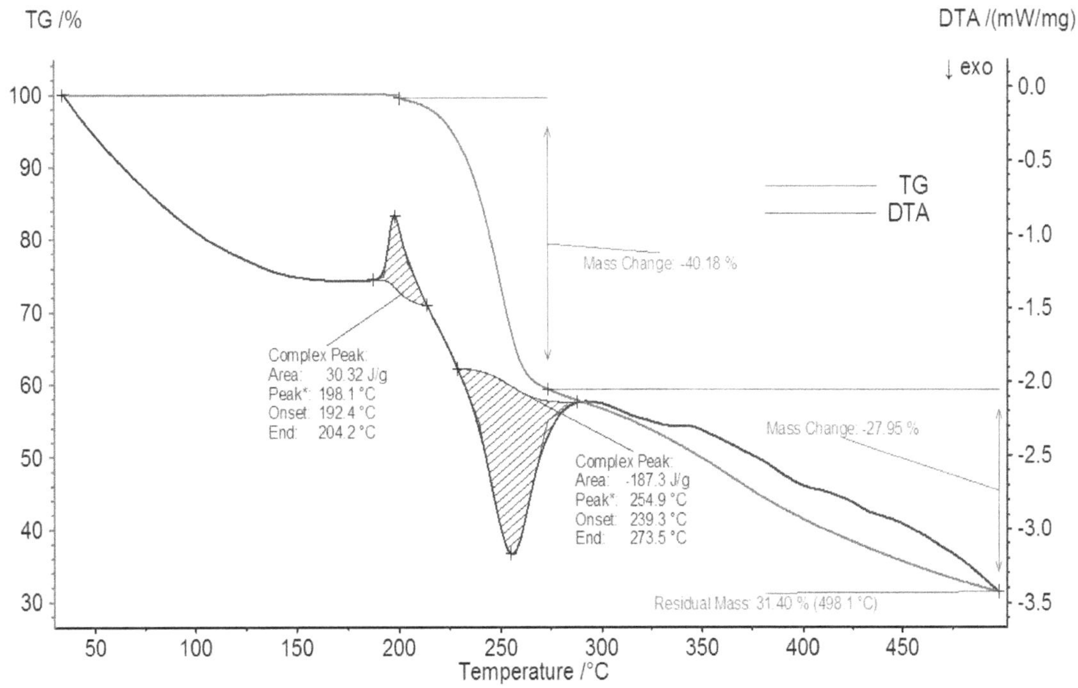


**Figure 5.3 TG-DTA Thermogram of DPGN**

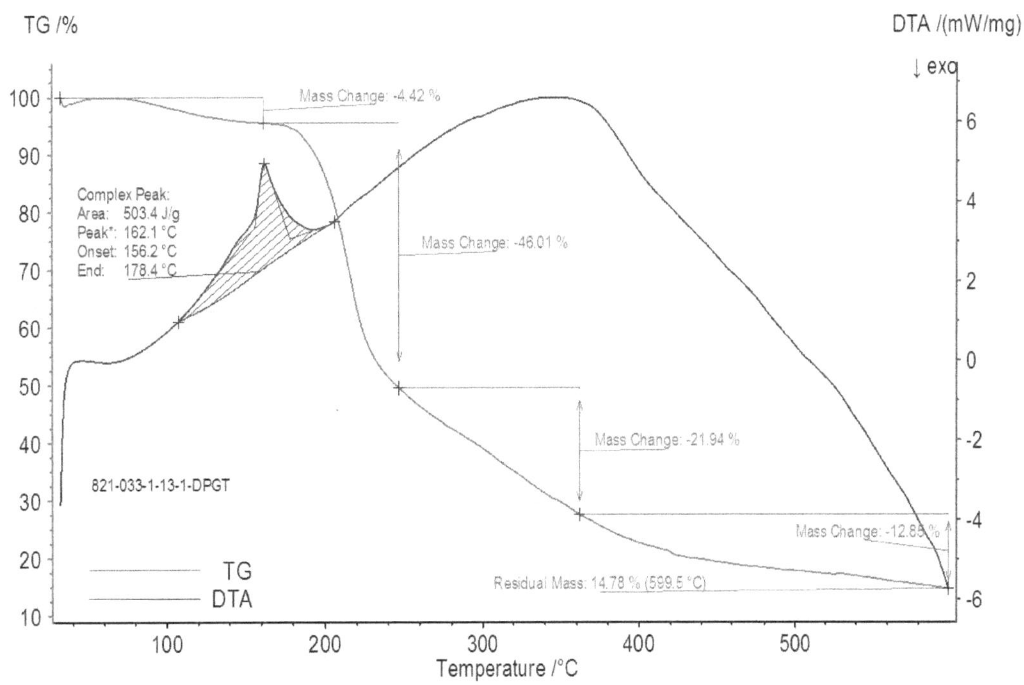


**Figure 5.4 TG-DTA Thermogram of DPGTM**

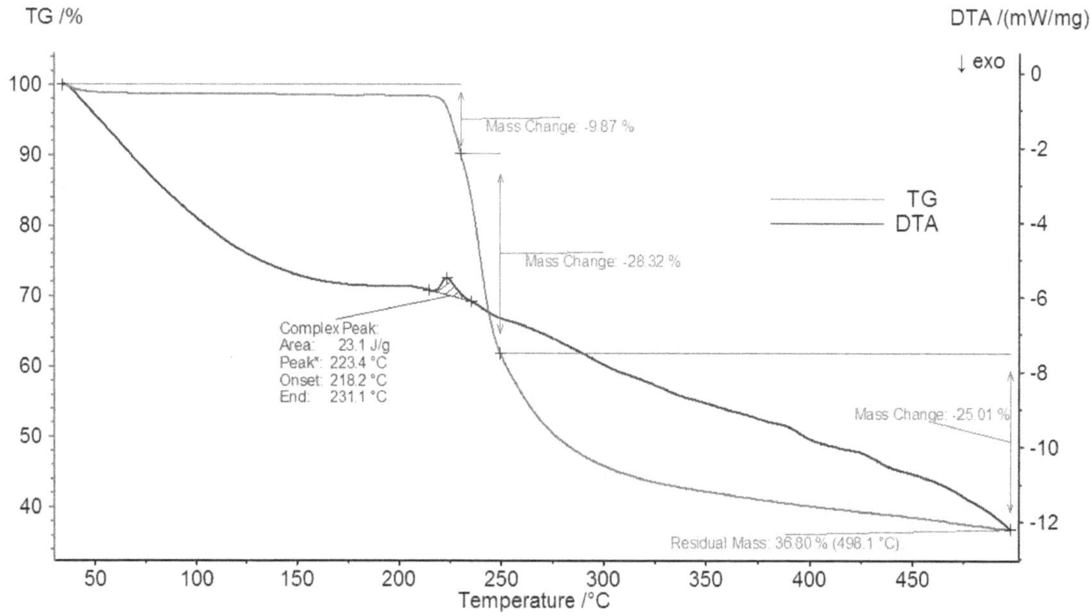


**Figure 5.5 TG-DTA Thermogram of DPGP**

## 5.4 MICROHARDNESS ANALYSIS

The measurement of hardness is a non-destructive testing method to determine the mechanical behavior of the materials. The hardness of a material is a measure of its resistance to plastic deformation which plays a vital role in device fabrication. By definition the indentation hardness is the ratio of the applied load to surface area generated to the indentation. As the hardness properties are basically related to the crystal structure and the bond strength of the material, micro hardness studies have been applied to understand the plasticity of the crystals (Westbrook and Conrod, 1971). Hardness tests are commonly carried out to determine the mechanical strength of materials and it correlates with other mechanical properties, like elastic constants and yield strength (Taber, 1951). Hardness measurements can be defined as macro, micro and nano according to the forces applied and displacement obtained (Pethica and Taber, 1979; Pethica et. al., 1983).

### 5.4.1 Various Tests on Micro Hardness

Hardness measurements can be carried out by various methods, they are:

(i). static indentation test,

(ii). damping (or) dynamic indentation test,

(iii). scratch test,

(iv). rebound test and

(v). abrasion test.

The popular and simplest among the above tests is static indentation test, in which a steady load is applied to an indentor. The indentor may be a ball or diamond cone or diamond pyramid. The hardness is calculated from the area or the depth of indentation produced. The variables are the type of indentor or load. The indentor is made up of a very hard material to prevent its deformation by the test piece, so that it can cover materials over a wide range of hardness. For this reason either a hardened steel sphere or a cone or a diamond pyramid is employed. In general, pyramid indentor is preferred to obtain geometrically similar impression for various loads. In the process of static indentation test, the indentor is pressed perpendicularly into the surface of a sample by means of an applied load. By measuring the depth of indentation and knowing the applied load, the hardness number may be evaluated. This procedure is followed in Brinell, Mayer, Vicker's, Knoop and Rockwell's tests. In dynamic indentation test, a ball or a cone is allowed to fall from a definite height and the hardness number is calculated from the dimensions of the indentation and the energy of impact. The scratch test can be classified into two types:

(i). comparison test, in which one material is said to be harder than another, if the second material is scratched by the first and

(ii). groove test, in this type of scratch test a diamond or steel indention is made on the surface at a steady rate under a definite load. The hardness number is expressed in terms of the width or depth of the groove formed.

In rebound test, an object of standard mass and dimension is allowed to bounce from the test surface. The height of rebound is taken as the measure of hardness. In abrasion test, a specimen is loaded against a rotating disk and the rate of wear is taken as a measure of hardness.

### 5.4.2 Vicker's Micro Hardness Test

Among the various types of micro hardnesses measurements, the most common and reliable one is the Vicker's micro hardness test. It uses a square diamond pyramid of two pairs of opposite faces with included angle of 136° each and all the four faces meet at a point. The geometry of Vicker's test is shown in figure 5.6. Smith and Sandland, (1922) have proposed this test.

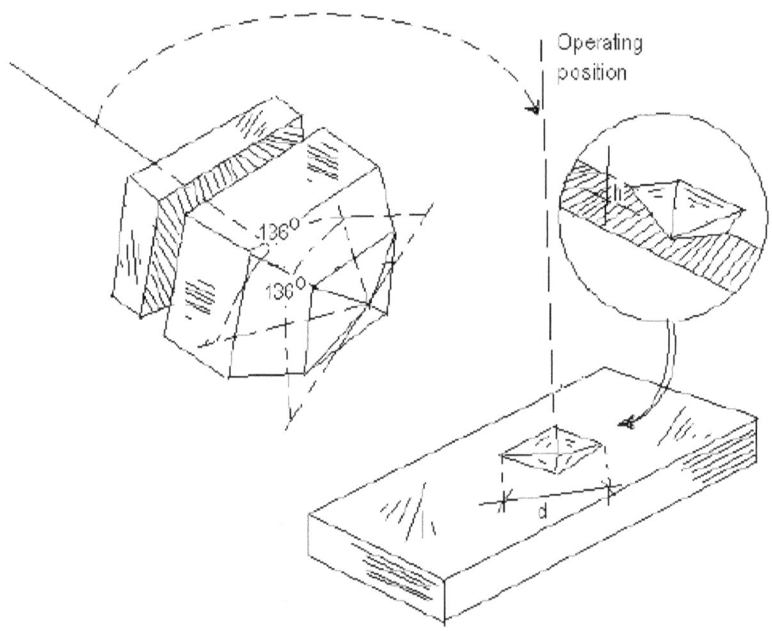


**Figure 5.6 Geometry of Vicker's diamond pyramidal indentor**

A pyramidal indentor is best suited for hardness tests due to the following reasons (Balta et. al., 1980);

(i). the contact pressure for a pyramidal indentor is independent of indentor size and elastic release corrections are less in pyramidal indentor.

(ii). the hardness value is independent of depth of penetration, since all indentations are geometrically similar.

The base of the Vicker's pyramid is a square and the depth of indentation corresponds to 1/17th of the indentation diagonal. The hardness, $H_V$ is the ratio between the applied load (kg) and the actual area of impression ($mm^2$). The Vicker's hardness number ($H_V$) is expressed as,

$$H_V = \frac{2PSin\,(\theta/2)}{d^2} \qquad (5.4)$$

where 'θ' the apex angle of the indentor ($\theta = 136^o$) and d the average diagonal length of the indentation impression (mm). Thus $H_V$ in $kg/mm^2$ is given by

$$H_V = \frac{1{,}8544\,P}{d^2} \qquad (5.5)$$

### 5.4.3 Experiment and Theory

Interpretation of hardness is perceived as ability of material to resist permanent deformation (Mott, 1956). Selected smooth and flat surfaces of the grown crystal was subjected to micro hardness study at room temperature with different load using Vicker's hardness tester (LEITZ WETZLER) fitted with a diamond pyramidal indenter and attached to an optical microscope. The indentation time was kept constant for all the loads. The Vicker's hardness number was calculated using the relation given in equation (5.5). The relationship between load and size of the indentation is given by the well-known Meyer's law:

$$P = ad^n \qquad (5.6)$$

where the exponent 'n' called as the Meyer's number or work hardening coefficient, and 'a' the constant (Sangwal et. al., 2003).

The work hardening coefficient (n) is calculated from the plot of *log P* versus *log d* of the grown crystal. The phenomenon of dependence of micro hardness of a solid on the applied load at low level of testing load is known as indentation size effect (ISE). The decrease in hardness with increase in load is usually termed as normal ISE whereas the increase in hardness with increase in load is called as reverse ISE.

The simplest way to describe the ISE is Meyer's law. For the normal ISE behavior, the exponent $n < 2$ but $n > 2$ for the reverse ISE behavior. When $n = 2$, the hardness is independent of the applied test load, and is given by Kick's law (Russel Raj and Murugakoothan, 2012).

### 5.4.4 Mechanical Analysis of N,N'-Diphenylguanidine Based NLO Single Crystals

Micro hardness study of any system has direct correlation with the crystal structure and is very sensitive to the presence of other phases or phase transition prevalent in the system (Viswanath et. al., 2007). The hardness measurements for DPGN, DPGTM and DPGP crystals were carried out on their prominent (010), (100) and (110) planes respectively. The micro hardness value was calculated using the relation (5.5). Indentations were made on the (010), (100) and (110) faces of the crystal. The micro hardness measurements were made for the applied loads varying from 5 to 50 g by keeping the constant dwell time of 3 s for all cases. A plot between the hardness number and load for the grown crystals are shown in figure 5.7.

We clearly infer from the figure 5.7 that the micro hardness number increases with increasing the load. This trend is in agreement with the reverse indentation size effect (RISE) (Onitsch, 1947). The maximum value of hardness number observed for DPGN, DPGTM and DPGP crystals are 61.84 kg/mm$^2$ (at 30 g), 54.04 kg/mm$^2$ (at 30 g), 36.96 kg/mm$^2$ (at 50 g) respectively. The load above these values develops multiple cracks on the crystal surface due to the release of internal stresses generated locally by indentation.

A plot obtained between log (P) and log (d), are shown in figure 5.8, gives a straight line. The value of work hardening coefficient (n) obtained for DPGN, DPGTM and DPGP crystals on their prominent planes using linear fit is found to be 2.1, 3.9 and 2.6 respectively. It is evident from the above plot that the micro hardness value of the crystal increases with increase in load which is in agreement with the reverse indentation size effect (RISE). According to Onitsch (1956), the work hardening coefficient ($n$) lies between 1 and 1.6 for hard materials and is more than 1.6 for soft materials. The value of '$n$' obtained for the grown crystals, shown in figure 5.8, are in the range 2.1 – 3.9 which indicate that the grown crystals belong to soft material category.

An elastic stiffness constant or simply an elastic constant measures how “hard” a solid is. An elastic stiffness constant ($C_{11}$) for the title compounds were calculated using the Wooster’s empirical formula $C_{11} = H_v^{7/4}$. The plots of stiffness constant versus load P are shown in figure 5.9. The trend of stiffness constant graph is similar to hardness number graph which gives an idea about tightness of bonding between neighboring atoms (Wooster, 1953). The high value of elastic stiffness constants for different loads for the grown crystals shows that atoms are tightly bonded and are depicted in Table 5.3.

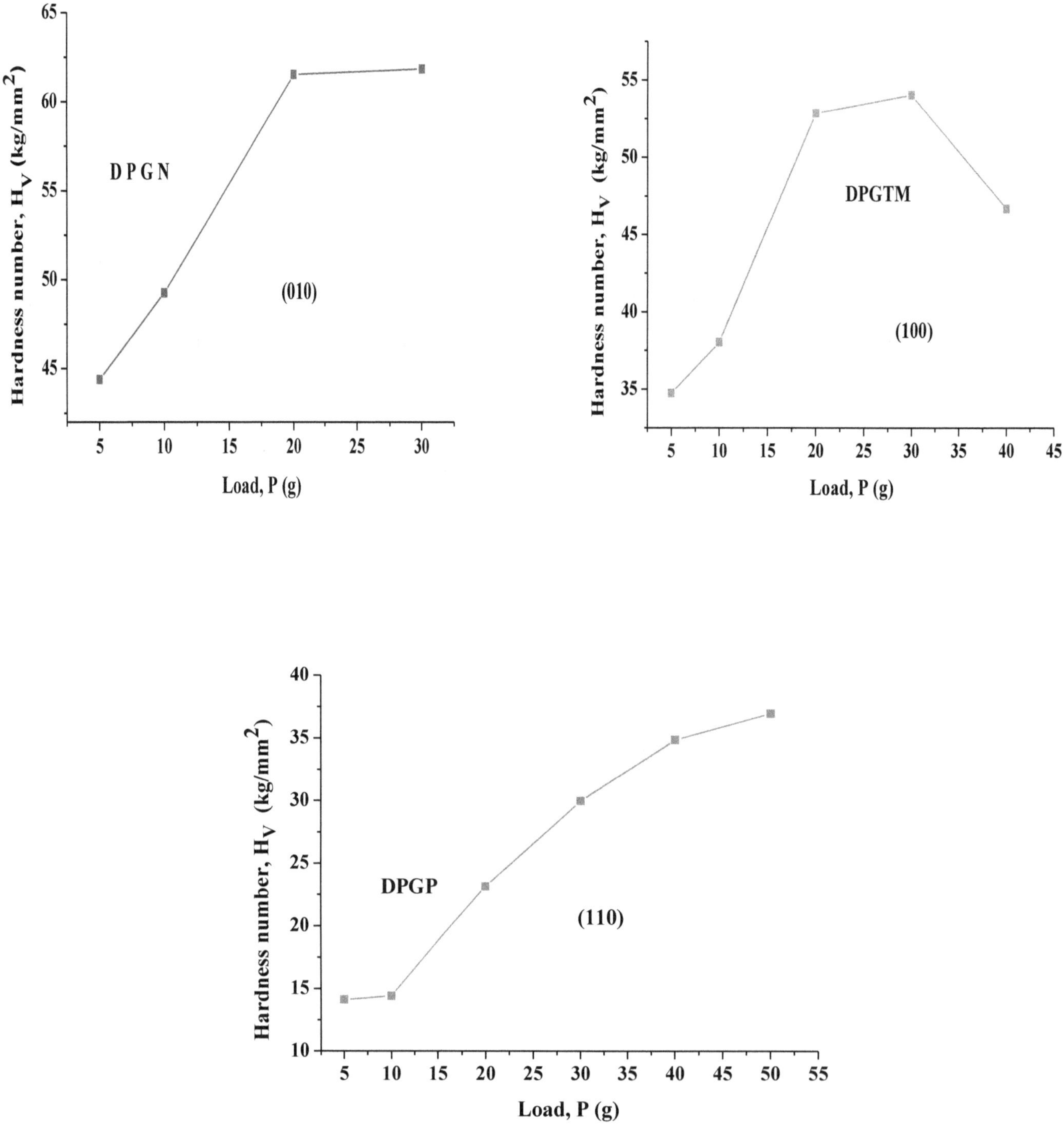


**Figure 5.7 Hardness number vs load for DPGN, DPGTM and DPGP single crystals**

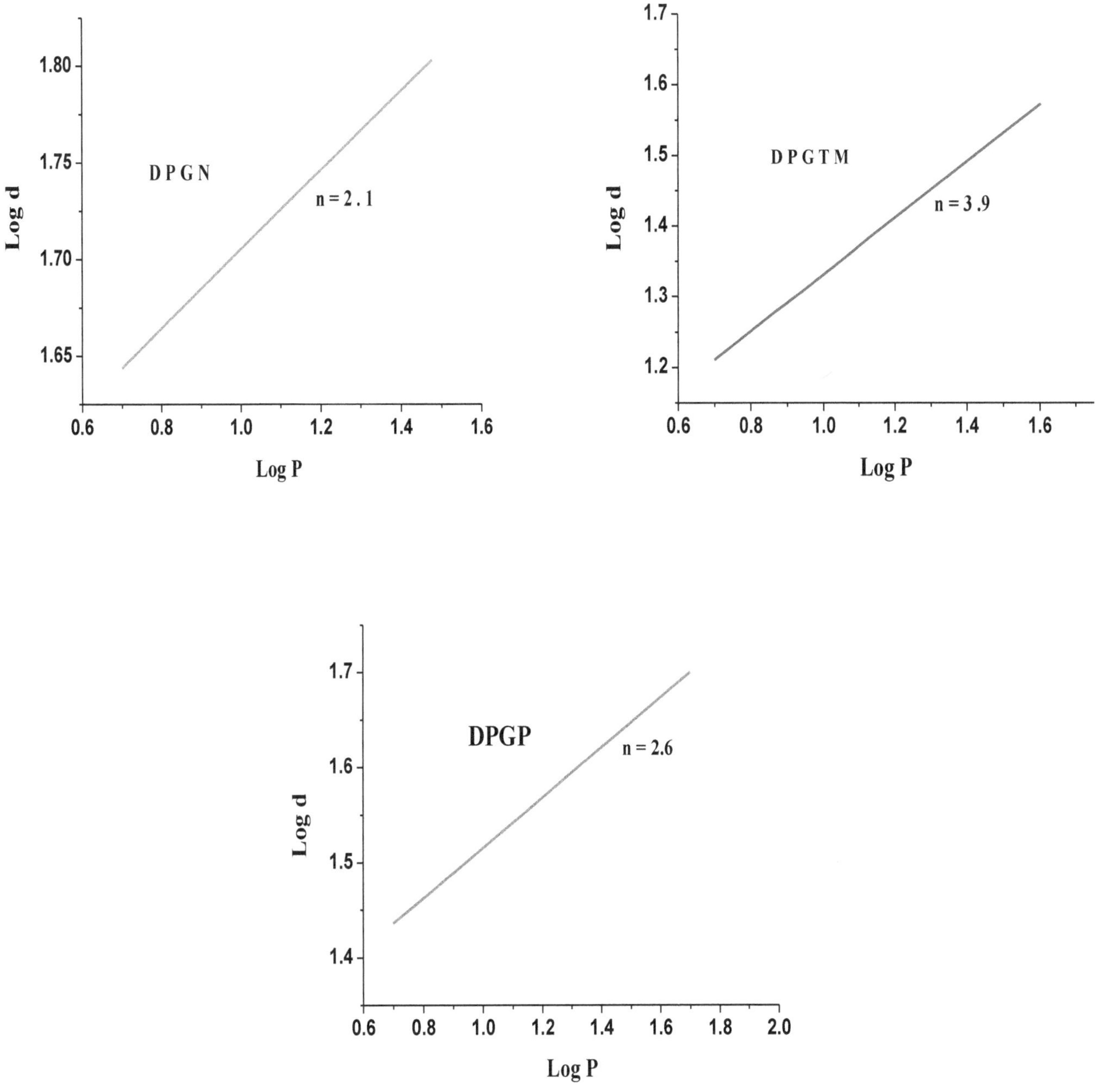


**Figure 5.8 Plot of Log P versus Log d for DPGN, DPGTM and DPGP single crystals**

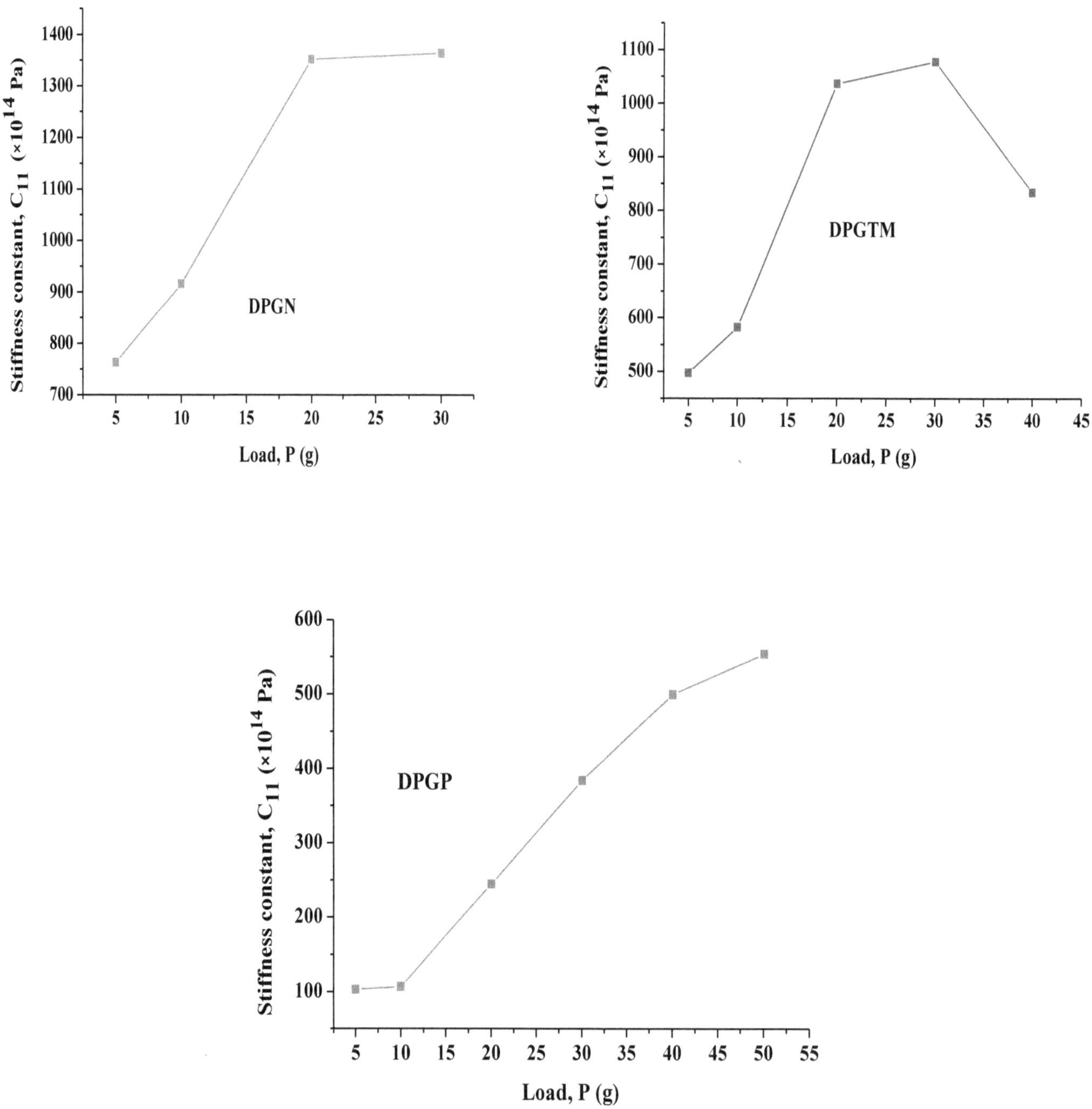


**Figure 5.9 Plot of Load versus stiffness constant for DPGN, DPGTM and DPGP single crystals**

**Table 5.3 Values of elastic stiffness constants for different loads for the grown crystals**

| Load (g) | $C_{11}$ ($10^{14}$ Pa) | | |
|---|---|---|---|
| | DPGN | DPGTM | DPGP |
| 5 | 763.09 | 497.10 | 102.97 |
| 10 | 915.61 | 582.12 | 106.44 |
| 20 | 1351.76 | 1036.26 | 244.50 |
| 30 | 1363.70 | 1077.08 | 383.88 |
| 40 | - | 833.32 | 499.86 |
| 50 | - | - | 554.02 |

## 5.5 CONCLUSION

The dielectric properties of the grown crystals were studied, and the different types of polarizations present in the grown crystals were confirmed by the dielectric studies. High purity of the grown crystals was confirmed from the lower dielectric loss values. The thermal stability of the grown crystals were studied by TG-DTA analysis. The N,N' - diphenylguanidine based compounds, studied in this work, such as DPGN, DPGTM and DPGP are found to decompose before melting at 192 °C, 100 °C and 224 °C respectively. Mechanical properties of the grown crystals were determined by Vicker's micro-indentation studies, which shows that these crystals are soft and mechanically stable. It appears that the single crystals of DPGN, DPGTM and DPGP are satisfying reverse indentation size effect. Owing to the better electrical, thermal and mechanical behaviors the title crystals studied in the present work can be handled easily during the fabrication of high quality optical devices and they can withstand in the working temperature region.

# CHAPTER - VI

# SUMMARY AND SUGGESTIONS FOR FUTURE WORK

## 6.1 SUMMARY

The increasing quest for variety of applications of non-linear optical (NLO) materials for second harmonic generation (SHG), sum or difference frequency mixing, optical parametric oscillation, amplification etc., have resulted in the development of numerous organic, inorganic and semi-organic NLO crystals. Nonlinear optical crystals should meet several requirements, such as large phase-matching angle, large nonlinear optical coefficient, a wide optical window around the visible region, good mechanical and chemical stability and a high laser damage threshold. Organic NLO materials offer several advantages, such as large nonlinear figure of merit, fast response time and their structural flexibility facilitates easy control of physical properties over very wide range. But their moderate environmental stability, low mechanical strength and a limited temperature range of operation limits their application. Inorganic materials possess good mechanical and thermal stabilities, however the optical nonlinearity is relatively lower than organic materials. The growth of inorganic crystals is usually more difficult and typically adopts high temperature growth technique.

In order to overcome the drawbacks of both organic and inorganic materials a new class of material comprising the advantages of both organic and inorganic materials was developed. Semi-organic NLO crystals seem to be very promising among the three categories which share the advantages of both organic and inorganic. As a result, semi-organic complexes have come under considerable scrutiny. The relatively simple guanidinium cation and its derivatives forms a broad family of hydrogen-bonded crystals. Guanidine compounds and its family derivatives seem to be easy to obtain and is cheaper than complicated amino acids. This is relevant for the applications of guanidinium based materials to tailor the optical properties for nonlinear optical devices. In general, guanidinium compounds are very stable and have good NLO properties.

In the present thesis, the growth, characterization and theoretical investigations of organic N,N'- diphenylguanidinium hydrogen (+) – L – tartrate monohydrate (DPGTM) and semi-organic materials such as, N,N'- diphenylguanidinium nitrate (DPGN), N,N'- diphenylguanidinium dihydrogen phosphite (DPGP) crystals were dealt in detail.

Optically good quality single crystals, such as N,N'- diphenylguanidinium nitrate (DPGN), N,N'- diphenylguanidinium hydrogen (+)-L-tartrate monohydrate (DPGTM) and N,N'- diphenylguanidinium dihydrogen phosphite (DPGP) crystals were grown by slow evaporation technique. Solvents used for the synthesis and growth are methanol and mixed solvents of water + ethanol (1:1). The solubility of title compounds was determined for their corresponding solvents at different temperatures and it indicates that the solubility increases linearly with temperature. The morphology of the grown crystals were analyzed in order to reveal their habit.

The grown crystals were analysed by single crystal X-ray diffraction (SXRD) and powder X-ray diffraction (PXRD) to confirm their structure and crystalline nature. The DPGN and DPGTM crystals belongs to orthorhombic crystal system with non-centrosymmetric space group $Pna2_1$ and $P2_12_12_1$. The DPGP crystallize in tetragonal crystal system with non-centrosymmetric space group $P4_3$. These parameters agrees well with that of the reported literature. The PXRD study revealed that the title compounds are crystalline in nature. The DFT was employed to optimize the structural geometry of the title compounds. B3LYP/ 6-31G (d,p) level basis set was used to obtain the ground state geometry for DPGN and DPGP. B3LYP/ 6-31G (d) level of basis set was used to obtain optimized geometry for DPGTM. The optimized molecular geometry was found to be true without any negative frequencies. The obtained geometrical parameters were compared with that of the experimental data and they are found to be in good agreement. FT-IR and FT-

Raman spectral analyses were performed to study the molecular vibrations and functional groups for the grown crystals. These spectral studies confirm the formation of the title compounds.

The grown crystals of thickness 2 mm were used for UV-vis transmittance analysis. From the transmittance spectra, the percentage of transmission, the transparency range and the band gap energy of the grown crystals were determined. It is also observed that the experimental band gap of the grown crystals are in the range between 3.90 – 4.14 eV, thereby favoring optical applications. The HOMO – LUMO band gap were computed and found to be comparable with the experimental values. From HOMO – LUMO energy gap, the chemical hardness of the title compounds was calculated and found to be chemically stable and thereby paving way for physical hardness. The existence of second harmonic generation in the title compounds was assessed by the Kurtz-Perry powder technique. It can be concluded that all the title compounds are NLO active and their nonlinearity is almost equivalent to that of KDP. The nonlinear optical parameters, such as dipole moment, polarizability and first order hyperpolarizability were computed in order to understand NLO behaviour at molecular level. The nonlinear optical parameters of title compounds were found to be non-zero and ahead of KDP. The laser damage threshold study confirm that the grown crystals exhibit higher stability to laser and high value of LDT compared with KDP.

The dielectric properties of the grown crystals were studied, and the different types of polarizations present in the grown crystals were confirmed by the dielectric studies. High purity of the grown crystals was confirmed from the lower dielectric loss values. The thermal stability of the grown crystals was studied by TG-DTA analysis. The N,N'- diphenylguanidine based compounds, studied in this work, such as, DPGN, DPGTM and DPGP are found to decompose before melting at 192 °C, 100 °C and 224 °C respectively. Mechanical properties of the grown crystals were determined by Vicker's micro-indentation studies, which shows that these crystals are soft and mechanically stable. It appears that the single crystals of DPGN, DPGTM and DPGP are satisfying reverse indentation size effect.

From the promising characteristics, such as excellent optical transparency, good second harmonic generation nonlinearity, high laser damage threshold, high mechanical and thermal stability the grown N,N'-diphenylguanidine based single crystals are found to be the potential candidates for nonlinear optical and optoelectronics device fabrication.

## 6.2 SUGGESTIONS FOR FUTURE WORK

In the present investigation the title crystals were grown by slow evaporation technique. Attempt could be made to grow bulk crystals by slow cooling method by optimizing growth parameters and also by studying the nucleation parameters for device applications. Attempts could be made to adopt Sankaranarayanan-Ramasamy (SR) method to grow bulk single crystals of N,N'-diphenylguanidine series along different growth axis in order to investigate their growth rate, transparency and nonlinear optical properties. Also, attempt could be made on to improve its transparency and SHG nonlinearity by adding some metal ions and rare earth metal ions as dopants. The etching behavior on the different growth planes can be done using various organic solvents in order to estimate the dislocation density. From these studies, the application oriented plane of the crystals could be identified and focused for device fabrications.

The particle size dependent SHG studies can be carried out on the title compounds and dielectric tensor analysis will also be carried out on these crystals. The Z-scan, Maker-fringes, phase matching and birefringence analysis may be performed on these crystals. In theoretical aspects the simulated structure of these crystals can be used to study other parameters, such as potential energy scan (PES), Mulliken charge analysis, molecular electrostatic potential (MEP) and thermodynamic properties etc.

# REFERENCES


1. Aakeroy C.B., Hitchcock P.B., Moyle B.D. and Seddon K.R., (1993a), 'Crystal engineering: hydrogen-bonded salts of hydroxybenzoic acids for second harmonic generation', J. Chem. Soc., Chem. Commun., Vol. 2, pp.152-156.

2. Aakeröy C.B., and Seddon K.R., (1993b), 'The hydrogen bond and crystal engineering', Chem. Soc. Rev., Vol.22, pp.397-407.

3. Anandha Babu G., Sreedhar S., Venugopal Rao S., and Ramasamy P., (2010), 'Synthesis, growth, structural, thermal, linear and nonlinear optical properties of a new organic crystal: Dimethyl ammonium picrate', J. Cryst. Growth, Vol. 312, pp. 1957–1962.

4. Antony C.J., Junaid Bushiri M., Hema Tresa Varghese, Yohannan Panicker C., and Michel Fleck., (2009), 'Spectroscopic properties of guanidinium zinc sulphate $[C(NH_2)_3]_2Zn(SO_4)_2$ and ab initio calculations of $[C(NH_2)_3]_2$ and $HC(NH_2)_3$', Spectrochim. Acta Part A., Vol. 73, pp. 942–945.

5. Arumanayagam T., and Murugakoothan P., (2012), 'Investigation on Nonlinear optical guanidinium based single crystals, Ph.D., Thesis.

6. Arumanayagam T., and Murugakoothan P., (2013), 'Studies on growth, spectral and mechanical properties of new organic NLO crystal: Guanidinium 4-nitrobenzoate (GuNB)', J. Cryst. Growth, Vol. 362, pp. 304–307.

7. Ashour A., El – Kadry N., and Mahmud S.A., (1995), 'On the electrical and optical properties of CdS films thermally deposited by a modified source', Thin Solid Films, Vol.269, pp.117-120.

8. Baccaro S., Cecilia A., Montecchi M., Malatesta T., Notaristefani F., Torrioli S., and Vittori F., (1998), 'Refractive index and absorption length of YAP : Ce scintillation crystal and reaectance of the coating used in YAP : Ce single-crystal matrix', Nucl. Instr. and Meth. in Phys. Res. A., Vol. 406, pp. 479-485.

9. Bailey R.T., Cruickshank F.R., Pavlides P., Pugh D., and Sherwood J.N., (1991), 'Organic materials for non-linear optics: inter-relationships between molecular properties, crystal structure and optical properties', J. Phys. D, Vol.24, pp.135-145.

10. Balta Calleja F.J., Rueda D.R., and Mead W.T., (1980), 'New aspects of the microhardness of ultraorientated polyethylene', J. Mater. Sci., Vol. 15, pp. 765–772.

11. Boling N.L., Crisp M.D., and Dube G., (1973), 'Laser Induced Surface Damage', Appl. Opt., Vol.12, pp.650-660.

12. Boyd R.W., (2007), 'Nonlinear optics', Third Edition, Academic Press, New York.

13. Brice J. C., (1972), 'The growth of crystals from liquids', Wiley, New York.

14. Brice J.C., (1986), 'Crystal Growth Processes', Halsted Press, John Wiley and Sons, New York.

15. Buckley H.E., (1951), 'Crystal growth', John Wiley and Sons, New York.

16. Chemla D.S., and Zyss J., (1987), 'Nonlinear Optical Properties of Organic Molecules and Crystals', Academic Press, New York.

17. Chen C., Wu Y., Jiang A., Wu B., You G., Li R., and Lin S., (1989), 'New nonlinear-optical crystal: $LiB_3O_5$', J. Opt. Soc. Am. B., Vol. 6 (4), pp.616-621.

18. Christian R., (1990), 'Solvents and solvent effects in organic chemistry', VCH, New York.

19. Colthup N.B., Daly L.H., and Wiberely S.E., (1964), 'Introduction to Infrared & Raman Spectroscopy', Academic Press, New York.

20. Davis E.A., and Mott N.F., (1970), 'Conduction in non-crystalline systems V. Conductivity, optical absorption and photoconductivity in amorphous semiconductors', Philosophical Magazine, Vol.22, pp.903 – 922.

21. Debrus S., Ratajczak H., Venturini J., Pincon N., Baran J., Barycki J., Glowiak T., and Pietraszko A., (2002), 'Novel nonlinear optical crystals of noncentrosymmetric structure based on hydrogen bonds interactions between organic and inorganic molecules', Synth. Mat., Vol.127, pp.99 – 104.

22. Dewey C.F., Cook W.R., Hodgson R.T., and Wynne J.J., (1975), 'Frequency doubling in $KB_5O_8.4H_2O$ and $NH_4B_5O_8.4H_2O$ to 217.3 nm', Appl. Phys. Lett., Vol.26, pp.714-716.

23. Douglas A.S., James H. F., and Timothy A.N., (2004), 'Principles of Instrumental Analysis', Fourth edu. Books/Cole.

24. Drozd M., Baran J., and Pietraszko A., (2005a), 'Crystal structure, differential scanning calorimetry and vibrational low temperature investigation of C(NH2)3·HSeO4', Spectrochim. Acta Part A., Vol. 61, pp. 2775-2787.

25. Drozd M., and Baran J., (2005b), 'Polarized IR-microscope spectra of guanidinium hydrogen selenate single crystal', Spectrochim. Acta Part A., Vol. 61, pp. 2953–2965.

26. Drozd M., Baran J., and Pietraszko A., (2005c), 'Diguanidinium hydrogenarsenate monohydrate and its deuterated analogue vibrational, DSC and X – ray investigations', Spectrochim. Acta Part A., Vol.61, pp. 2809 - 2821.

27. Dyke S.F., Floyd J., Sainsbury M., and Theobald R.S., (1978), 'Organic Spectroscopy: An Introduction', Longman, London.

28. Elleuch N., Amamoub W., Ben Ahmed A., Abid Y., and Feki H., (2014), 'Vibrational spectroscopic study, charge transfer interaction and nonlinear optical properties of L-asparaginium picrate: A density functional theoretical approach', Spectrochim. Acta Part A., Vol. 128, pp. 781–789.

29. Elwell D., and Scheel H. J., (1975), 'Crystal growth from high temperature solution', Academic press, Inc. London.

30. Eugene G.T., Galina G.K., and Vladimir P.S., (2005), 'General approaches to design of a reproducible technique for the growth of large crystals of barium metaborate (BBO) for industrial application', J. Cryst. Growth, Vol. 275, pp. e2123–e2128.

31. Foresman J.B., and Frisch A., (1993), 'Exploring Chemistry with Electronic Structure Methods', Gaussian Inc., Pittsburgh.

32. Frisch, M. J.; Trucks, G. W.; Schlegel, H. B.; Scuseria, G. E.; Robb, M. A.; Cheeseman, J. R.; Montgomery, Jr., J. A.; Vreven, T.; Kudin, K. N.; Burant, J. C.; Millam, J. M.; Iyengar, S. S.; Tomasi, J.; Barone, V.; Mennucci, B.; Cossi, M.; Scalmani, G.; Rega, N.; Petersson, G. A.; Nakatsuji, H.; Hada, M.; Ehara, M.; Toyota, K.; Fukuda, R.; Hasegawa, J.; Ishida, M.; Nakajima, T.; Honda, Y.; Kitao, O.; Nakai, H.; Klene, M.; Li, X.; Knox, J. E.; Hratchian, H. P.; Cross, J. B.; Bakken, V.; Adamo, C.; Jaramillo, J.; Gomperts, R.; Stratmann, R. E.; Yazyev, O.; Austin, A. J.; Cammi, R.; Pomelli, C.; Ochterski, J. W.; Ayala, P. Y.; Morokuma, K.; Voth, G. A.; Salvador, P.; Dannenberg, J. J.; Zakrzewski, V. G.; Dapprich, S.; Daniels, A. D.; Strain, M. C.; Farkas, O.; Malick, D. K.; Rabuck, A. D.; Raghavachari, K.; Foresman, J. B.; Ortiz, J. V.; Cui, Q.; Baboul, A. G.; Clifford, S.; Cioslowski, J.; Stefanov, B. B.; Liu, G.; Liashenko, A.; Piskorz, P.; Komaromi, I.; Martin, R. L.; Fox, D. J.; Keith, T.; Al-Laham, M. A.; Peng, C. Y.; Nanayakkara, A.; Challacombe, M.; Gill, P. M. W.; Johnson, B.; Chen, W.; Wong, M. W.; Gonzalez, C.; and Pople, J. A.; Gaussian 03 Revision C.02, Gaussian Inc., Wallingford CT, 2004.

33. Gupta S.S., Pradhan R.D., Marcano A., Melikechi N., and Desai C.F., (2002), 'Laser damage studies in zinc (tris) thiourea sulfate: A nonlinear optical crystal', J. Appl. Phys., Vol. 91, pp. 3125-3128.

34. Halasyamani P.S., and Poeppelmeier K.R., (1998), 'Non-centrosymmetric oxides', Chem. Mater., Vol.10, pp.2753-2769.

35. Hopper R.M., Naranes R.S., Hearole B.J. and Sherwood J.N., (1980),'Crystal Growth Second Edition', Ed. Pamplin B.R., Pergamon Press, New York.

36. Hu Q., Jin H. L., Chen X. A., and Wang S., (2012), 'Thermal and FT-IR spectral studies of N,N'-diphenylguanidine', J. Therm. Anal. Calorim., Vol. 110, pp. 593 – 599.

37. James C., Jayakumar V.S., and Hubert Joe I., (2007), 'Natural bond orbital analysis and vibrational spectroscopic studies of H-bonded N,N'-diphenylguanidinium nitrate', J. Mol. Str., Vol 830, pp. 156-166.

38. Jiang X.N., Xu D., Yuan D.R., Lu M.K., and Sun D.L., (2002), 'Distinct Growth Phenomena Observed on Zinc Cadmium Thiocyanate Crystals by Atomic Force Microscopy', Cryst. Res. Technol., Vol.6, pp. 564 - 569.

39. Johnson B.G., Gill P.M.W., and Pople J.A., (1993), 'The performance of a family of density functional methods', J. Chem. Phys., Vol. 98, pp.5612-5626.

40. Kar T., (2012), 'Amino acids-precursors for synthesizing nonlinear optical materials', Prog. Cryst. Growth Characterization, Vol.58, pp.74-83.

41. Kleinman D.A., (1962), 'Nonlinear Dielectric Polarization in Optical Media', Phy. Rev., Vol. 126, pp. 1977–1979.

42. Kohn W., and Sham L.J., (1965), 'Self-Consistent Equations Including Exchange and Correlations Effects', Phys. Rev., Vol. 140, pp. A1133 – A1138.

43. Koopmans T.A., (1933), Physica, Vol. 1, pp. 104–113.

44. Krishnan S., Raj C.J., Dinakaran S. and Das S.J., (2008), 'Investigation of optical band gap in potassium acid phthalate single crystal', Cryst. Res. Technol., Vol. 43, pp. 670-673.

45. Krol H., Gallais L., Grezes-Besset C., Natoli J.Y., and Commandre M., (2005), 'Investigation of nanoprecursors threshold distribution in laser damage testing', Optic. Comm., Vol. 256, pp. 184-189.

46. Kurtz S.K. and Perry T.T. (1968), 'A powder technique for the evaluation of nonlinear optical materials', J. Appl. Phys., Vol. 39, pp. 3798–3813.

47. Laudise R.A., (1970), 'The growth of single crystals', prentice hall, Eaglewood Cliffs, New Jersey.

48. Lee C., Yang W., and Parr R.G., (1988), 'Development of the Colle-Salvetti correlation-energy formula into a functional of the electron density', Phys. Rev. B., Vol. 37, 785-789.

49. Lee W.T., Salje E.K.H. and Martin T D., (1999), 'Effect of surface relaxations on the equilibrium growth morphology of crystals: platelet formation', J. Phys. Condens. Matter, Vol.11, pp.7385-7410.

50. Levine I.N., (1991), 'Quantum Chemistry', $4^{th}$ ed., Prentice-Hall, Inc., Englewood Cliffs, New Jersey.

51. Majchrowski A., Borowiec M. T. and Michalski E., (2004), 'Top seeded solution growth of KHo $(WO_4)_2$ single crystals', J. Cryst. Growth, Vol. 264, pp. 201–207.

52. Makhatadze G.I., (1999), 'Thermodynamics of protein interactions with urea and guanidinium hydrochloride', J. Phys. Chem. B, Vol.103, pp. 4781– 4785.

53. Marek Szafranski and Andrzej Katrusiak (2006), 'Effect of halogen atom exchange on the thermodynamic behavior and ferroelectric properties of $[C(NH_2)_3]_4 Br_2SO_4$', Phys. Rev. B., Vol. 73, pp. 134111-1 – 8.

54. Meirs H.A., and Issac F., (1987), 'The spontaneous crystallisations', Proc. Roy. Soc., Vol. A79, pp.322-325.

55. Mott B.W., (1956), 'Microindentation Hardness Testing', Butterworths, London.

56. Mullin J.W (1972), 'Crystallization', Second edition, Academic press, London.

57. Mullin. J.W, (1976), 'Industrial crystallization 78', Plenum Press, New York.

58. Nabil E., Ali B.A., Habib F., Younes A., and Christian M., (2014), 'Vibrational spectra, optical properties, NBO and HOMO–LUMO analysis of L-Phenylalanine L-Phenylalaninium Perchlorate: DFT calculations', Spectrochim. Acta Part A., Vol. 121, pp. 129–138.

59. Nakamoto K., (1986), 'Infrared and Raman Spectra of Inorganic and Coordination Compounds, Wiley, New York.

60. Nalwa H.S., and Miyata. S., (1997a), Editors, 'Organic Electroluminescent Materials and Devices', Amsterdam: Gordon and Breach.

61. Nalwa H.H., Nalwa H.S., and Miyata S., (1997b), 'Nonlinear Optics of Organic Molecules and Polymers, CRC Press Inc., New York.

62. Narayana Bhat M., and Dharmaprakash S. M., (2002), 'Growth of nonlinear optical $\gamma$ -glycine crystals', J. Cryst. Growth, Vol. 236, pp. 376–380.

63. Nyvlt I., Rychly R., and Gottried Wutzelova J., (1970), 'Metastable zone width of some aqueous solutions' J. Cryst. Growth, Vol.6, pp.151-162.

64. Oleg Kolosov, Alexei Gruverman, Jun Hatano, Koichiro Takahashi, and Hiroshi Tokumoto, (1995), 'Nanoscale Visualization and Control of Ferroelectric Domains by Atomic Force Microscopy', Phys. Rev. Lett., Vol. 74, pp. 4309-4315.

65. Onitsch E.M., (1947), 'Über die Mikrohärte der Metalle', Mikroskopie, Vol.2, pp.131.

66. Onitsch E.M., (1956), 'The present status of testing hardness of materials', Microscopie, Vol. 95, pp. 12–14.

67. Paixao J.A., Pereira Silva P.S., Matos Beja A., Ramos Silva M., and Alte da Veiga L., (1998), 'Crystal structure of N,N' – Diphenylguanidinium Nitrate', Acta Cryst. C., Vol. 54, pp. 805-808.

68. Paixao J.A., Pereira Silva P.S., Matos Beja A., Ramos Silva M., De Matos Gomes E and Belsley M., (1999), 'Crystal structure of N,N' – Diphenylguanidinium hydrogen (+) – L – Tartrate monohydrate', Acta Cryst. C., Vol. 55, pp. 1287-1290.

69. Paixao J.A., Matos Beja A., Ramos Silva M., and Alte da Veiga L., (2001), 'Crystal structure of N,N' – Diphenylguanidinium dihydrogen phosphite', Z.Kristallogr. NCS., Vol. 216, pp. 416 - 418.

70. Pamplin B.R., (1979), 'Crystal Growth', Pergamon Press, Oxford.

71. Parr R.G., Donnelly R.A., Levy M., and Palke W.E., (1978), 'Electronegativity: The density functional viewpoint', J. Chem. Phys., Vol. 68, pp. 3801 – 3807.

72. Parr R.G., and Yang W., (1989), 'Density Functional Theory of Atom and Molecules', Oxford University Press, New York.

73. Pearson R.G., (1989), 'Absolute electronegativity and hardness: applications to organic chemistry', J. Org. Chem., Vol. 54, pp. 1423–1430.

74. Pearson R.G., (1997), 'Chemical Hardness-Applications from Molecules to Solids', John Wiley –VCH, Weinheim.

75. Pethica J.B., and Taber D., (1979), 'Contact characterized metal surface at very low loads: deformation and adhesion', Surf. Sci., Vol.89, pp. 182–190.

76. Pethica J.B., Huchingts R., and Oliver W.C., (1983), 'Hardness measurements at penetration depths as small as 20 nm', Philos. Mag. (A), Vol. 48, pp. 593–606.

77. Ponpandian N., Balaya P., and Narayanasamy A., (2002), 'Electrical Conductivity and Dielectric Behaviour of Nanocrystalline NiFe2O4 Spinel', J. Phys. Condens. Matter, Vol.14, pp. 3221-3237.

78. Pople J.A., and Beveridge D.L., (1970), 'Approximate Molecular Orbital Theory', McGraw-Hill: New York.

79. Prabha K., Rajesh Kumar T., Shi Feng Du., Vimalan M., Dayalan A., and Sagayaraj P., (2010), 'Growth and characterization of a new nonlinear optical barium strontium borate (BSB) single crystal', Mater. Chem. Phys., Vol. 121, pp. 22-27.

80. Prasad P.N., and Williams D.J., (1991), 'Introduction to Nonlinear Optical Effects in Molecules and Polymers', John Wiley & Sons, Inc. New York.

81. Rak M., Eremin N.N., Eremina T.A., Kuznetsov V.A., Ohkrimenko T.M., Furmanova N.G., and Efremova E.P., (2005), 'On the mechanism of impurity influence on growth kinetics and surface morphology of KDP crystals—I: defect centres formed by bivalent and trivalent impurity ions incorporated in KDP structure - theoretical study', J. Cryst. Growth, Vol.273, pp.577 - 585.

82. Russel Raj K., and Murugakoothan P., (2012), 'Studies on the optical and mechanical properties of non-linear optical 3-aminophenol orthophosphoric acid (3-amphph) single crystal', Optik, Vol. 123, pp. 1082–1086.

83. Russel Raj K., and Murugakoothan P., (2013), 'Growth and physical properties of a new crystal for NLO applications: Bisguanidinium hydrogen phosphate monohydrate (G2HP)', J. Cryst. Growth., Vol. 362, pp. 130–134.

84. Sajan D., Hubert J., Jayakumar V.S., Zaleski J., (2006) , 'Structural and electronic contibutions to hyperpolarizability in methyl p-hydroxybenzoate', J. Mol. Struct., Vol. 785, pp. 43–53.

85. Santhanaragavan P., and Ramasamy P., (2000), 'Crystal growth methods and process', Kuru publications, Kumbakonam.

86. Sangwal K., Surowska B., and Blaziak P., (2003), 'Relationship between indentation size effect and material properties in the microhardness measurement of some cobalt-based alloys', Mater. Chem. Phys., Vol. 80 (2), pp. 428–437.

87. Saravana Kumar G., and Murugakoothan P., (2014), 'Synthesis, spectral analysis, optical and thermal properties of new organic NLO crystal: N,N'-Diphenylguanidinium Nitrate (DPGN), Spectrochim. Acta Part A., Vol. 131, pp. 17–21.

88. Saravana Kumar G., and Murugakoothan P., (2015a), 'Synthesis, spectral, optical and electric studies of N,N'-diphenylguanidinium hydrogen (+)-l-tartrate monohydrate – A new organic NLO crystal, Optik, Vol. 126, pp. 68–73.

89. Saravana Kumar G., Mano Balaji G., and Murugakoothan P., (2015b), 'Experimental and theoretical investigations on N,N'-diphenylguanidinium dihydrogen phosphite – A semi-organic nonlinear optical material, Spectrochim. Acta Part A., Vol. 138, pp. 340 – 347.

90. Sauter E.G., (1996), 'Nonlinear Optics', Wiley, New York.

91. Science News Online (6/6/1998), 'Melanoma Madness The scientific flap over sunscreens and skin cancer - Chemical studies'.

92. Siva shankar V., Sankar R., Siddheswaran R., Jayavel R. and Murugakoothan P. (2008), 'Growth and characterization of tetra l-lysine alanine mono hydrochloride dihydrate (TLAMHCl), a new semiorganic nonlinear optical single crystal', Mat. Chem. Phys., Vol. 109, pp. 119-124.

93. Smith R.L., and Sandland G., (1922), 'An accurate method of determining hardness of metal with reference to those of a high degree of hardness', Proc. Inst. Mech. Enggrs., Vol. 102, pp. 623–641.

94. Smith B.C., (1999), 'Infrared Spectral interpretations: A systematic Approach', CRC press, LLC.

95. Stephens P.J., Devlin F.J., Chablowski C.R., and Frisch M.J., (1994), 'Ab Initio calculation of vibrational absorption and circular dichrosism spectra using density functional force fields', J. Phys. Chem., Vol. 98 (45), pp.11623-11627.

96. Stuart B.C., Feit M.D., Rubenchik A.M., Shore B.W., and Perry M.D., (1995), 'Laser-Induced Damage in Dielectrics with Nanosecond to Subpicosecond Pulses', Phys. Rev. Lett., Vol. 74, pp. 2248-2251.

97. Sudharsana N., Keerthana B., Nagalakshmi R., Krishnakumar V., and Guru Prasad L., (2012), 'Growth and characterization of hydroxyethyl ammonium picrate single crystals for third-order nonlinear optical applications', Mater. Chem. Phys., Vol.134, pp. 736–746.

98. Suvitha A., and Murugakoothan P., (2012), 'Synthesis, growth, structural, spectroscopic and optical studies of a semiorganic NLO crystal: Zinc guanidinium phosphate', Spectrochim. Acta Part A., Vol. 86, pp. 266 – 270.

99. Suvitha A., Vivek P., and Murugakoothan P., (2013), 'Nucleation kinetics. Growth and characteriz,ation of guanidinium 3-nitrobenzoate single crystal', Optik., Vol. 124, pp. 3534– 3538.

100. Taber D., (1951), 'The hardness of materials', Clarendon Press, Oxford.

101. Tauc J., Grigorovici R., and Vancu A., (1966), 'Optical Properties and Electronic Structure of Amorphous Germanium', Phys. Status Solid, Phys. Status Solidi B., Vol.15, pp.627 – 637.

102. Usharani S., (2000), 'Analytical Chemistry', Macmillan publication, India Ltd.

103. Von Hippel P.H., and Schleich T.W., (1969), 'Ion effects on the solution structure of biological macromolecules'. Acc. Chem. Res., Vol. 2, pp. 257-265.

104. Vosko S.H., Wilk I., and Nusair M., (1980), 'Accurate spin-dependent electron liquid correlation energies for local spin density calculations: a critical analysis, Can. J. Phys., Vol. 58 (8), pp. 1200 - 1211.

105. Vijayan N., Bhagavannarayana G., Gopalakrishnan R., and Ramasamy P., (2007), 'Structural and optical characterization on solution grown methyl p-hydroxybenzoate single crystals', Indian J. Chem., Sect. A, Vol. 46, pp. 70-73.

106. Viswanath B., Raghavan R., Ramamurty U., and Ravishankara N., (2007), 'Mechanical properties and anisotropy in hydroxyapatite single crystals', Scripta Mater., Vol. 57, pp. 361 – 364.

107. Vivek P., and Murugakoothan P., (2013), 'Growth and anisotropic studies on potential nonlinear optical crystal imidazole–imidazolium picrate monohydrate (IIP) in different orientations for NLO device fabrications', Opt. & Laser tech., Vol. 49, pp. 288 – 295.

108. Warren B.E., (1990), 'X-ray Diffraction', Dover Publications, Mineola, New York.

109. Wendlandt W.W., Kasper M., and Bellamy S., (1984), 'A TG-DSC Investigation of the thermal dissociation of selected guanidinium salts', Thermochim. Acta, Vol. 15, pp. 239 – 244.

110. Westbrook J.H., and Conrod H., (1971), 'The science hardness testing and research applications', American society for metals, Ohio.

111. William H.Z., (2004), 'Theory of X-ray Diffraction in Crystals', Edition Dover Phoenix, Dover Publications, Mineola, New York.

112. Wooster W. A., (1953), 'Physical properties and atomic arrangements in crystals', Reports on Progress in Physics, Vol 16, pp. 62–82.

113. Yang W., Parr R.G., and Uytterhoeven L., (1987), 'New relation between hardness and compressibility of minerals, Phys. Chem. Minerals., Vol 15(2), pp. 191 – 195.

114. Zhang K.C., and Zhang L.H., (1997), 'Science and Technology of Crystal Growth', Science Press, Beijing.

115. Zyss J., Pecaut J., Levy J.P. and Masse R., (1993), 'Synthesis and Crystal Structure of Guanidinium L-Monohydrogentartrate: Encapsulation of an Optically Nonlinear Octupolar Cation', Acta Cryst., Vol. B49, pp. 334-342.

116. Zyss J., (1994), 'Molecular Nonlinear Optics: Materials, Physics, and Devices', Academic Press, San Diego, California.

www.ingramcontent.com/pod-product-compliance
Lightning Source LLC
LaVergne TN
LVHW082341200926
848481LV00012B/3413
* 9 7 9 8 2 3 2 0 9 1 3 8 5 *